Universitext

Series Editors:

Sheldon Axler
San Francisco State University

Vincenzo Capasso
Università degli Studi di Milano

Carles Casacuberta
Universitat de Barcelona

Angus J. MacIntyre
Queen Mary, University of London

Kenneth Ribet
University of California, Berkeley

Claude Sabbah
CNRS, École Polytechnique

Endre Süli
University of Oxford

Wojbor A. Woyczynski
Case Western Reserve University

Universitext is a series of textbooks that presents material from a wide variety of mathematical disciplines at master's level and beyond. The books, often well class-tested by their author, may have an informal, personal even experimental approach to their subject matter. Some of the most successful and established books in the series have evolved through several editions, always following the evolution of teaching curricula, to very polished texts.

Thus as research topics trickle down into graduate-level teaching, first textbooks written for new, cutting-edge courses may make their way into Universitext.

For further volumes:
http://www.springer.com/series/223

Universitext

Universitext is a series of textbooks that presents material from a wide variety of mathematical disciplines at master's level and beyond. The books, often well class-tested by their author, may have an informal, personal even experimental approach to their subject matter. Some of the most successful and established books in the series have evolved through several editions, always following the evolution of teaching curricula, to very polished texts.

Thus as research topics trickle down into graduate-level teaching, first textbooks written for new, cutting-edge courses may make their way into *Universitext*.

For further volumes:
http://www.springer.com/series/223

Rodney Coleman

Calculus on Normed
Vector Spaces

Springer

Rodney Coleman
Laboratoire Jean Kuntzmann
Grenoble
France

ISSN 0172-5939   ISSN 2191-6675 (electronic)
ISBN 978-1-4614-3893-9   ISBN 978-1-4614-3894-6 (eBook)
DOI 10.1007/978-1-4614-3894-6
Springer New York Heidelberg Dordrecht London

Library of Congress Control Number: 2012939881

Mathematics Subject Classification (2010): 46–XX, 46N10

Printed on acid-free paper

Springer is part of Springer Science+Business Media (www.springer.com)

To Francie, Guillaume, Samuel
and Jeremy

Preface

The aim of this book is to present an introduction to calculus on normed vector spaces at a higher undergraduate or beginning graduate level. The prerequisites are basic calculus and linear algebra. However, a certain mathematical maturity is also desirable. All the necessary topology and functional analysis is introduced where necessary.

I have tried to show how calculus on normed vector spaces extends the basic calculus of functions of several variables. I feel that this is often not done and we have, on the one hand, very elementary texts, and on the other, high level texts, but few bridging the gap.

In the text there are many nontrivial applications of the theory. Also, I have endeavoured to give exercises which seem, at least to me, interesting. In my experience, very often the exercises in books are trivial or very academic and it is difficult to see where the interest lies (if there is any!).

In writing this text I have been influenced and helped by many other works on the subject and by others close to it. In fact, there are too many to mention; however, I would like to acknowledge my debt to the authors of these works. I would also like to express my thanks to Mohamed El Methni and Sylvain Meignen, who carefully read the text and gave me many helpful suggestions.

Writing this book has allowed me to clarify many of my ideas and it is my sincere hope that this work will prove useful in aiding others.

Grenoble, France   Rodney Coleman

Contents

Chapter 1
Normed Vector Spaces

In this chapter we will introduce normed vector spaces and study some of their elementary properties.

1.1 First Notions

We will suppose that all vector spaces are real. Let E be a vector space. A mapping $\| \cdot \| : E \longrightarrow \mathbb{R}$ is said to be a *norm* if, for all $x, y \in E$ and $\lambda \in \mathbb{R}$ we have

- $\|x\| \geq 0$;
- $\|x\| = 0 \Longleftrightarrow x = 0$;
- $\|\lambda x\| = |\lambda| \|x\|$;
- $\|x + y\| \leq \|x\| + \|y\|$.

The pair $(E, \| \cdot \|)$ is called a *normed vector space* and we say that $\|x\|$ is the norm of x. The fourth property is often referred to as the normed vector space *triangle inequality*.

Exercise 1.1. Show that $\| - x \| = \|x\|$ and that $\| \cdot \|$ is a convex function, i.e.,

$$\|\lambda x + (1 - \lambda)y\| \leq \lambda \|x\| + (1 - \lambda)\|y\|$$

for all $x, y \in E$ and $\lambda \in [0, 1]$.

When there is no confusion, we will simply write E for a normed vector space. To distinguish norms on different normed vector spaces we will often use suffixes. For example, if we are dealing with the normed vector spaces E and F, we may write $\| \cdot \|_E$ for the norm on E and $\| \cdot \|_F$ for the norm on F. The most common norms on $\mathbb{R}^n$ are defined as follows:

$$\|x\|_1 = |x_1| + \cdots + |x_n|, \quad \|x\|_2 = \sqrt{x_1^2 + \cdots + x_n^2} \quad \text{and}$$

$$\|x\|_\infty = \max\{|x_1|, \ldots, |x_n|\},$$

R. Coleman, *Calculus on Normed Vector Spaces*, Universitext,
DOI 10.1007/978-1-4614-3894-6_1, © Springer Science+Business Media New York 2012

where x_i is the ith coordinate of x. There is no difficulty in seeing that $\| \cdot \|_1$ and $\| \cdot \|_\infty$ are norms. For $\| \cdot \|_2$ the only difficulty can be found in the last property. If we set $x \cdot y = \sum_{i=1}^{n} x_i y_i$, the dot product of x and y, and write $\| \cdot \|$ for $\| \cdot \|_2$, then

$$p(t) = t^2 \|x\|^2 + 2t(x \cdot y) + \|y\|^2 = \|tx + y\|^2 \geq 0.$$

As p is a second degree polynomial and always nonnegative, we have

$$4\big((x \cdot y)^2 - \|x\|^2 \|y\|^2\big) \leq 0$$

and so

$$\|x + y\|^2 = (x + y) \cdot (x + y) = \|x\|^2 + \|y\|^2 + 2(x \cdot y)$$
$$\leq \|x\|^2 + \|y\|^2 + 2\|x\|\|y\| = (\|x\| + \|y\|)^2.$$

This gives us the desired inequality.

If $n = 1$, then these three norms are the same, i.e.,

$$\|x\|_1 = \|x\|_2 = \|x\|_\infty = |x|.$$

Exercise 1.2. Characterize the norms defined on $\mathbb{R}$.

In general, we will suppose that the norm on $\mathbb{R}$ is the absolute value.

It is possible to generalize the norms on $\mathbb{R}^n$ defined above. We suppose that $p > 1$ and for $x \in \mathbb{R}^n$ we set $\|x\|_p = (\sum_{i=1}^{n} |x_i|^p)^{\frac{1}{p}}$.

Proposition 1.1. $\|x\|_p$ *is a norm on* $\mathbb{R}^n$.

Proof. It is clear that the first three properties of a norm are satisfied. To prove the triangle inequality, we proceed by steps. We first set $q = \frac{p}{p-1}$ and prove the following formula for strictly positive numbers a and b:

$$a^{\frac{1}{p}} b^{\frac{1}{q}} \leq \frac{a}{p} + \frac{b}{q}.$$

For $k \in (0, 1)$ let the function $f_k : \mathbb{R} \longrightarrow \mathbb{R}$ be defined by $f_k(t) = k(t-1) - t^k + 1$. Then $f_k(1) = 0$ and $\frac{df_k}{dt}(t) = k(t - t^{k-1})$. It follows that $f_k(t) \geq 0$ if $t \geq 1$ and so, for $k \in (0, 1)$ and $t \geq 1$,

$$t^k \leq tk + (1 - k).$$

If we set $t = \frac{a}{b}$ and $k = \frac{1}{p}$ when $a \geq b$, or $t = \frac{b}{a}$ and $k = \frac{1}{q}$ when $a < b$, then we obtain the result.

The next step is to take $x, y \in \mathbb{R}^n \setminus \{0\}$ and set $a = (\frac{|x_i|}{\|x\|_p})^p$ and $b = (\frac{|y_i|}{\|y\|_q})^q$ in the formula. We obtain

$$\frac{|x_i y_i|}{\|x\|_p \|y\|_q} \le \frac{1}{p} \left(\frac{|x_i|}{\|x\|_p} \right)^p + \frac{1}{q} \left(\frac{|y_i|}{\|y\|_q} \right)^q,$$

and then, after summing over i,

$$\sum_{i=1}^{n} |x_i||y_i| \le \|x\|_p \|y\|_q.$$

This inequality clearly also holds when $x = 0$ or $y = 0$.

Now,

$$\|x + y\|_p^p \le \sum_{i=1}^{n} |x_i||x_i + y_i|^{p-1} + \sum_{i=1}^{n} |y_i||x_i + y_i|^{p-1}$$

and, using the inequality which we have just proved,

$$\sum_{i=1}^{n} |x_i||x_i + y_i|^{p-1} \le \|x\|_p \left(\sum_{i=1}^{n} |x_i + y_i|^{(p-1)q} \right)^{\frac{1}{q}}$$

$$= \|x\|_p \left(\sum_{i=1}^{n} |x_i + y_i|^p \right)^{\frac{1}{q}}$$

$$= \|x\|_p \|x + y\|_p^{\frac{p}{q}}.$$

In the same way

$$\sum_{i=1}^{n} |y_i||x_i + y_i|^{p-1} \le \|y\|_p \|x + y\|_p^{\frac{p}{q}}$$

and so

$$\|x + y\|_p^p \le (\|x\|_p + \|y\|_p) \|x + y\|_p^{\frac{p}{q}},$$

from which we obtain the triangle inequality. $\qquad\square$

Exercise 1.3. Show that

$$\lim_{p \to 1} \|x\|_p = \|x\|_1 \qquad \text{and} \qquad \lim_{p \to \infty} \|x\|_p = \|x\|_\infty$$

for any $x \in \mathbb{R}^n$.

Let E be a vector space and $\mathcal{N}(E)$ the collection of norms defined on E. We define a relation $\sim$ on $\mathcal{N}(E)$ by writing $\|\cdot\| \sim \|\cdot\|^\times$ if there exist constants $\alpha > 0$ and $\beta > 0$ such that

$$\alpha \|x\|^\times \le \|x\| \le \beta \|x\|^\times$$

for all $x \in E$. This relation is an equivalence relation. If $\| \cdot \| \sim \| \cdot \|^{\times}$, then we say that the two norms are *equivalent*.

Exercise 1.4. Establish the inequalities $\|x\|_{\infty} \leq \|x\|_2 \leq \|x\|_1 \leq n\|x\|_{\infty}$ and deduce that these three norms on $\mathbb{R}^n$ are equivalent.

If S is a nonempty set, then a real-valued function d defined on the Cartesian product S^2 is said to be a *metric* (or *distance*) if it satisfies the following properties for all $x, y, z \in S^2$:

- $d(x, y) \geq 0$;
- $d(x, y) - 0 \longleftrightarrow x = y$;
- $d(x, y) = d(y, x)$;
- $d(x, y) \leq d(x, z) + d(z, y)$.

We say that the pair (S, d) is a *metric space* and that $d(x, y)$ is the *distance* from x to y. The fourth property is referred to as the (metric space) triangle inequality. If $(E, \| \cdot \|)$ is a normed vector space, then it easy to see that, if we set $d(x, y) = \|x - y\|$, then d defines a metric on E. Many of the ideas in this chapter can be easily generalized to general metric spaces.

Exercise 1.5. What is the distance from $A = (1, 1)$ to $B = (4, 5)$ for the norms we have defined on $\mathbb{R}^2$?

Consider a point a belonging to the normed vector space E. If $r > 0$, then the set

$$B(a, r) = \{x \in E : \|a - x\| < r\}$$

is called the *open ball* of centre a and radius r. For $r \geq 0$ the set

$$\bar{B}(a, r) = \{x \in E : \|a - x\| \leq r\}$$

is called the *closed ball* of centre a and radius r. In $\mathbb{R}$ the ball $B(a, r)$ (resp. $\bar{B}(a, r)$) is the open (resp. closed) interval of length $2r$ and centre a. We usually refer to balls in the plane $\mathbb{R}^2$ as *discs*.

Exercise 1.6. What is the form of the ball $\bar{B}(0, 1) \subset \mathbb{R}^2$ for the norms $\| \cdot \|_1, \| \cdot \|_2$, and $\| \cdot \|_{\infty}$? (Notice that a ball may have corners.)

1.2   Limits and Continuity

We now consider sequences in normed vector spaces. If $(x_n)_{n \in \mathbb{N}}$ is a sequence in a normed vector space E and there is an element $l \in E$ such that $\lim_{n \to \infty} \|x_n - l\| = 0$, then we say that the sequence is *convergent*. It is easy to see that the element l must be unique. We call l the *limit* of the sequence and write $\lim_{n \to \infty} x_n = l$. We will in general abbreviate $(x_n)_{n \in \mathbb{N}}$ to (x_n) and $\lim_{n \to \infty} x_n = l$ to $\lim x_n = l$. The following result is elementary.

Proposition 1.2. *If* (x_n) *and* (y_n) *are convergent sequences in* E, *with* $\lim x_n = l_1$ *and* $\lim y_n = l_2$, *and* $\lambda \in \mathbb{R}$, *then*

$$\lim(x_n + y_n) = l_1 + l_2 \qquad \text{and} \qquad \lim(\lambda x_n) = \lambda l_1.$$

Suppose now that we have two normed vector spaces, $(E, \|\cdot\|_E)$ and $(F, \|\cdot\|_F)$. Let A be a subset of E, f a mapping of A into F and $a \in A$. We say that f is *continuous at* a if the following condition is satisfied:

for all $\epsilon > 0$, there exists $\delta > 0$ such that, if $x \in A$ and $\|x - a\|_E < \delta$,

then $\|f(x) - f(a)\| < \epsilon$.

If f is continuous at every point $a \in A$, then we say that f is *continuous* (on A). Finally, if $A \subset E$ and $B \subset F$ and $f : A \longrightarrow B$ is a continuous bijection such that the inverse mapping f^{-1} is also continuous, then we say that f is a *homeomorphism*.

Exercise 1.7. Suppose that $\|\cdot\|_E$ and $\|\cdot\|_E^\times$ are equivalent norms on E and $\|\cdot\|_F$ and $\|\cdot\|_F^\times$ equivalent norms on F. Show that f is continuous at a (resp. continuous) for the pair $(\|\cdot\|_E, \|\cdot\|_F)$ if and only if f is continuous at a (resp. continuous) for the pair $(\|\cdot\|_E^\times, \|\cdot\|_F^\times)$.

Proposition 1.3. *The norm on a normed vector space is a continuous function.*

Proof. We have

$$\|x\| = \|x - y + y\| \le \|x - y\| + \|y\| \implies \|x\| - \|y\| \le \|x - y\|.$$

In the same way $\|y\| - \|x\| \le \|y - x\|$. As $\|y - x\| = \|x - y\|$, we have

$$\big| \|x\| - \|y\| \big| \le \|x - y\|,$$

and hence the continuity. $\qquad\qquad\qquad\qquad\qquad\qquad\qquad\qquad\qquad\qquad\square$

The next result is also elementary.

Proposition 1.4. *Let* E *and* F *be normed vector spaces,* $A \subset E$, $a \in A$, f *and* g *mappings from* E *into* F *and* $\lambda \in \mathbb{R}$.

- *If* f *and* g *are continuous at* a, *then so is* $f + g$.
- *If* f *is continuous at* a, *then so is* λf.
- *If* α *is a real-valued function defined on* E *and both* f *and* α *are continuous at* a, *then so is* αf.

Corollary 1.1. *The functions* $f : E \longrightarrow F$ *which are continuous at* a *(resp. continuous) form a vector space.*

We now consider cartesian products of normed vector spaces. Let $(E_1, \| \cdot \|_{E_1}), \ldots, (E_p, \| \cdot \|_{E_p})$ be normed vector spaces. The cartesian product $E_1 \times \cdots \times E_p$ is a vector space. For $(x_1, \ldots, x_p) \in E_1 \times \cdots \times E_p$ we set

$$\|(x_1, \ldots, x_p)\|_S = \|x_1\|_{E_1} + \cdots + \|x_p\|_{E_p} \quad \text{and} \quad$$

$$\|(x_1, \ldots, x_p)\|_M = \max(\|x_1\|_{E_1}, \ldots, \|x_p\|_{E_p}).$$

There is no difficulty in seeing that $\| \cdot \|_S$ and $\| \cdot \|_M$ are equivalent norms on $E_1 \times \cdots \times E_p$. In general, we will use the second norm, which we will refer to as the usual norm. If $E_1 = \cdots = E_p = \mathbb{R}$ and $\| \cdot \|_{E_i} = | \cdot |$ for all i, then $\| \cdot \|_S = \| \cdot \|_1$ and $\| \cdot \|_M = \| \cdot \|_\infty$.

Proposition 1.5. *Let* $(E, \| \cdot \|)$ *be a normed vector space.*

- *The mapping* $f : E \times E \longrightarrow E, (x, y) \longmapsto x + y$ *is continuous.*
- *The mapping* $g : \mathbb{R} \times E \longrightarrow E, (\lambda, x) \longmapsto \lambda x$ *is continuous.*

Proof. Let us first consider f. We have

$$\|(x, y) - (a, b)\|_M < \epsilon \Longrightarrow \|x - a\| < \epsilon, \ \|y - b\| < \epsilon$$

$$\Longrightarrow \|(x + y) - (a + b)\| \leq \|x - a\| + \|y - b\| < 2\epsilon,$$

hence f is continuous at (a, b).

We now consider g. If $\|(\lambda, x) - (\alpha, a)\| < \epsilon$, then $|\lambda - \alpha| < \epsilon$ and $\|x - a\| < \epsilon$ and so

$$\|\lambda x - \alpha a\| = \|\lambda x - \lambda a + \lambda a - \alpha a\| \leq |\lambda| \|x - a\| + |\lambda - \alpha| \|a\| < (|\alpha| + \epsilon)\epsilon + \epsilon \|a\|,$$

therefore g is continuous at (α, a).   □

A composition of real-valued continuous functions of a real variable is continuous. We have a generalization of this result.

Proposition 1.6. *Let* E, F *and* G *be normed vector spaces,* $A \subset E$, $B \subset F$, f *a mapping of* A *into* F *and* g *a mapping of* B *into* G. *If* $f(A) \subset B$, f *is continuous at* $a \in A$ *and* g *continuous at* $f(a)$, *then* $g \circ f$ *is continuous at* a.

Proof. Let us take $\epsilon > 0$. As g is continuous at $f(a)$, there exists $\delta > 0$ such that, if $y \in B$ and $\|y - f(a)\|_F < \delta$, then $\|g(y) - g(f(a))\|_G < \epsilon$. As f is continuous at a, there exists $\alpha > 0$ such that, if $x \in A$ and $\|x - a\|_E < \alpha$, then $\|f(x) - f(a)\|_F < \delta$. This implies that $\|g(f(x)) - g(f(a))\|_G < \epsilon$. Therefore $g \circ f$ is continuous at a.
                                                                                          □

Corollary 1.2. *If* $A \subset E$ *and* $f : A \longrightarrow \mathbb{R}$ *is continuous and nonzero on* A, *then the function* $g = \frac{1}{f}$ *is continuous on* A.

Proof. It is sufficient to notice that g can be written $g = h \circ f$, where h is the real-valued function defined on R^* by $h(t) = \frac{1}{t}$. ◻

To close this section, we give a characterization of continuity which uses sequences.

Theorem 1.1. *Let E and F be normed vector spaces, $A \subset E$ and f a mapping of A into F. Then f is continuous at $a \in A$ if and only if, for every sequence (x_n) in A such that $\lim x_n = a$, we have $\lim f(x_n) = f(a)$.*

Proof. Suppose first that f is continuous at a and let us take $\epsilon > 0$. There exists $\delta > 0$ such that, if $x \in A$ and $\|x - a\|_E < \delta$, then $\|f(x) - f(a)\|_F < \epsilon$. Now let (x_n) be a sequence in A with limit a. There is an $n_0 \in \mathbb{N}$ such that, if $n \geq n_0$, then $\|x_n - a\|_E < \delta$. This implies that $\|f(x_n) - f(a)\|_F < \epsilon$. It follows that $\lim f(x_n) = f(a)$.

Now suppose that, when $(x_n) \subset A$ and $\lim x_n = a$, we have $\lim f(x_n) = f(a)$. If f is not continuous at a, then there is an $\epsilon > 0$ and a sequence $(x_n) \subset A$ such that $\|x_n - a\|_E < \frac{1}{n}$ and $\|f(x_n) - f(a)\|_F \geq \epsilon$. However, then $\lim x_n = a$ and the sequence $(f(x_n))$ does not converge to $f(a)$, a contradiction. So f must be continuous at a. ◻

1.3   Open and Closed Sets

Let E be a normed vector space. A subset O of E is said to be *open* if for every $x \in O$, there is an open ball centred on x which lies entirely in O. If $A \subset E$ and there is an open set O such that $a \in O \subset A$, then A is said to be a *neighbourhood* of a. If A is itself open, then we say that A is an *open neighbourhood* of a.

Proposition 1.7. *An open ball $B(a, r)$ is an open set.*

Proof. Let $x \in B(a, r)$ and set $\rho = \frac{1}{2}(r - \|x - a\|)$. Clearly $0 < \rho < r$ and, if $y \in B(x, \rho)$, then we have

$$\|a - y\| = \|a - x + x - y\| \leq \|a - x\| + \|x - y\| \leq \|a - x\| + \rho = r - \rho < r.$$

Therefore $B(x, \rho) \subset B(a, r)$ and it follows that $B(a, r)$ is open. ◻

If E is a normed vector space, then

- E and $\emptyset$ are open;
- if $(O_\alpha)_{\alpha \in A}$ is a collection of open subsets, then $\cup_{\alpha \in A} O_\alpha$ is an open set;
- if $(O_i)_{i=1}^n$ is a finite collection of open subsets, then $\cap_{i=1}^n O_i$ is an open set.

(If $\emptyset$ is not open, then there is an $x \in \emptyset$ such that for any $r > 0$ the ball $B(x, r) \not\subset \emptyset$. As there is no $x \in \emptyset$, this statement is false. It follows that $\emptyset$ is open.)

Exercise 1.8. Give an example of an infinite collection of open sets whose intersection is not open.

Notation. Let X be a set. We recall that, if S and T are subsets of X, then the *complement* of S in T is the subset of X

$$T \setminus S = \{x \in X : x \in T \text{ and } x \notin S\}.$$

We will refer to the set $X \setminus S$ simply as the complement of S and we will write cS for this subset. Clearly $T \setminus S = T \cap cS$.

If E is a normed vector space and $C \subset E$ is the complement of an open set, then we say that C is *closed*. Notice that a subset composed of a single point is a closed set. As might be expected, we have

Proposition 1.8. *A closed ball* $\bar{B}(a,r)$ *is a closed set.*

Proof. We need to show that the complement of $\bar{B}(a,r)$ is open. Let $x \in c\bar{B}(a,r)$ and $\rho = \frac{1}{2}(\|a - x\| - r)$. Clearly $\rho > 0$. If $y \in B(x,\rho)$, then

$$\|a - y\| \geq \|a - x\| - \|y - x\| > \|a - x\| - \rho = 2\rho + r - \rho = \rho + r > r.$$

Hence $B(x,\rho) \in c\bar{B}(a,r)$ and it follows that $c\bar{B}(a,r)$ is open.   □

Using de Morgan's laws, i.e.,

$$c(\cup_{\alpha \in A} A_\alpha) = \cap_{\alpha \in A} c A_\alpha \qquad \text{and} \qquad c(\cap_{\alpha \in A} A_\alpha) = \cup_{\alpha \in A} c A_\alpha,$$

we obtain: if a E is normed vector space, then

- E and $\emptyset$ are closed;
- if $(C_\alpha)_{\alpha \in A}$ is a collection of closed subsets, then $\cap_{\alpha \in A} C_\alpha$ is a closed set;
- if $(C_i)_{i=1}^n$ is a finite collection of closed subsets, then $\cup_{i=1}^n C_i$ is a closed set.

Exercise 1.9. Give an example of an infinite collection of closed sets whose union is not closed.

The union of all open sets included in a subset A of a normed vector space is called the *interior* of A and we write $\text{int}\,A$ for this set. The interior $\text{int}\,A$ is the largest open set lying in A. The intersection of all closed sets containing A is called the *closure* of A and we write $\bar{A}$ for this set. The closure $\bar{A}$ is the smallest closed set containing A. Clearly A is closed if and only if $A = \bar{A}$. The *boundary* of A, written ∂A, is the intersection of the sets $\bar{A}$ and $\overline{cA}$.

Exercise 1.10. Show that

- A is closed if and only if $\partial A \subset A$;
- $\bar{A}$ is the union of A and ∂A;
- $a \in \partial A$ if and only if every open neighbourhhood of a intersects both A and cA.

Exercise 1.11. Let A be a subset of a normed vector space. Show that, if O is a nonempty open subset of $\bar{A}$, then $O \cap A \neq \emptyset$.

Suppose now that $\| \cdot \|$ and $\| \cdot \|^\times$ are equivalent norms on a vector space E, i.e., there exist constants $\alpha > 0$ and $\beta > 0$ such that

$$\alpha \|x\|^\times \leq \|x\| \leq \beta \|x\|^\times$$

for all $x \in E$. Let us write $B(a, r)$ (resp. $B^\times(a, r)$) for the open ball of centre a and radius r defined with respect to the norm $\| \cdot \|$ (resp. $\| \cdot \|^\times$):

$$B(a,r) = \{x \in E : \|x - a\| < r\} \quad \text{and} \quad B^\times(a,r) = \{x \in E : \|x - a\|^\times < r\}.$$

Then

$$B^\times\left(a, \frac{r}{\beta}\right) \subset B(a, r) \subset B^\times\left(a, \frac{r}{\alpha}\right).$$

Hence $O \subset E$ is open for the norm $\| \cdot \|$ if and only if O is open for the norm $\| \cdot \|^\times$. This is also the case for closed subsets.

It is possible to characterize closed sets using sequences, as the next result shows.

Proposition 1.9. *Let E be a normed vector space and $A \subset E$. A is closed if and only if every convergent sequence in A has its limit in A.*

Proof. Suppose first that A is closed and let (x_n) be a convergent sequence contained in A. If $\lim x_n = x$, then every open ball centred on x contains an element of the sequence and hence an element of A. As cA is open, $x \notin cA$ and so $x \in A$.

Now consider the converse. Suppose that every convergent sequence included in A has its limit in A. Let $x \in cA$ and suppose that for any $n \in \mathbb{N}^*$ $B(x, \frac{1}{n})$ contains an element $x_n \in A$. The sequence (x_n) is included in A and converges to x and so, by hypothesis, $x \in A$. This however is a contradiction and so there exists an n such that $B(x, \frac{1}{n}) \subset cA$. It follows that cA is open and so A is closed. $\qquad\square$

In a normed vector space E the subsets E and $\emptyset$ are both open and closed. Are there any others? Suppose that $A \subset E$ is an open and closed subset and that $A \neq E$ and $A \neq \emptyset$. Also, let $a \in A$ and $b \in cA$ and let us set

$$\bar{t} = \sup\{t \in [0, 1] : a + s(b - a) \in A, 0 \leq s \leq t\}.$$

Clearly $\bar{t} \in (0, 1)$. Now we set $\bar{x} = a + \bar{t}(b - a)$ and $x_n = a + (\bar{t} - \frac{1}{n})(b - a)$ for $n \in \mathbb{N}^*$. The sequence (x_n) lies in A and converges to $\bar{x}$. As A is closed, $\bar{x}$ belongs to A. By definition of $\bar{t}$ there exists a sequence $(y_n) \subset cA$, where

$$y_n = a + (\bar{t} + \epsilon_n)(b - a) \quad \text{and} \quad 0 < \epsilon_n < \frac{1}{n}.$$

As $\lim y_n = \bar{x}$ and cA is closed, we have $\bar{x} \in cA$, which is a contradiction. Thus we have proved the

Proposition 1.10. *The only open and closed subsets of a normed vector space E are the sets E and $\emptyset$.*

The following result is a very useful characterization of continuous functions. It can also be used to decide whether a given set is open or closed.

Proposition 1.11. *Let E and F be normed vector spaces, $A \subset E$ and f a mapping from A into F. Then the following statements are equivalent:*

(a) *f is continuous;*
(b) *$f^{-1}(O)$ is the intersection of A with an open subset of E, if O is open in F;*
(c) *$f^{-1}(C)$ is the intersection of A with a closed subset of E, if C is closed in F.*

Proof. (a) $\Longrightarrow$ (b) Let $O \subset F$ be open. If $f(A) \cap O = \emptyset$, then $f^{-1}(O) = \emptyset$, an open subset of E; hence, the result is true in this case. Suppose now that $f(A) \cap O \neq \emptyset$ and let $f(a) \in O$. As O is open, there is an open ball $B(f(a), r) \subset O$. The continuity of f implies the existence of an open ball $B(a, \rho_a)$ such that $f(B(a, \rho_a) \cap A) \subset B(f(a), r)$. We obtain the inclusions

$$B(a, \rho_a) \cap A \subset f^{-1}(B(f(a), r)) \subset f^{-1}(O).$$

We now have

$$f^{-1}(O) = \cup(B(a, \rho_a) \cap A) = A \cap (\cup B(a, \rho_a)),$$

where the unions are taken over those $a \in A$ such that $f(a) \in O$. Thus $f^{-1}(O)$ is the intersection of A with an open subset in E.

(b) $\Longrightarrow$ (a) Let $a \in A$ and $r > 0$. As $B(f(a), r)$ is open, $f^{-1}(B(f(a), r))$ is the intersection of an open subset of E with A. However, $a \in f^{-1}(B(f(a), r))$, so there exists an open ball $B(a, \rho)$ whose intersection with A is contained in $f^{-1}(B(f(a), r))$. This inclusion implies that $f(B(a, \rho) \cap A) \subset B(f(a), r)$. It follows that f is continuous at a.

(b) $\Longrightarrow$ (c) If $C \subset F$ is closed, then cC is open and so $f^{-1}(cC)$ is the intersection of A with an open subset U of E. However,

$$f^{-1}(C) = A \setminus f^{-1}(cC) = A \cap cU$$

and so f is the intersection of A with a closed subset of E.

(c) $\Longrightarrow$ (b) This is proved in the same way as (b) $\Longrightarrow$ (c) $\square$

Exercise 1.12. Let E and F be normed vector spaces, $A \subset E$ and f a mapping from A into F. Show that if A is open (resp. closed) and f continuous, then $f^{-1}(B)$ is open (resp. closed), if B is an open (resp. closed) subset of F. Also, show that, if $f^{-1}(B)$ is open (resp. closed) when B is an open (resp. a closed) subset of F, then A is open (resp. closed) and f continuous. What can we say if $A = E$?

Example. Let $u : \mathbb{R}^n \longrightarrow \mathbb{R}$ be a continuous function and let us define $f : \mathbb{R}^n \times \mathbb{R} \longrightarrow \mathbb{R}$ by $f(x, y) = u(x) - y$. Then f is continuous and so $f^{-1}(0)$ is closed. However, $f^{-1}(0)$ is the graph of u; therefore, the graph of u is closed in $\mathbb{R}^n \times \mathbb{R}$.

1.4 Compactness

In this section we will introduce the fundamental notion of compactness. First we consider limit points. As above, we will suppose that E is a normed vector space. Let A be a subset of E. We say that $x \in E$ is a *limit point* of A if every open ball $B(x, r)$ contains points of A other than x. (Notice that x is not necessarily in A.) A is said to have the *Bolzano–Weierstrass property* if every infinite subset of A has a limit point in A.

Example. 0 is a limit point of the set $\{1, \frac{1}{2}, \frac{1}{3}, \ldots\}$. Are there any others?

Exercise 1.13. Show that $A \subset E$ is closed if and only if A contains all its limit points.

Suppose now that $\{O_i\}_{i \in I}$ is a collection of open subsets of E and $A \subset \cup_{i \in I} O_i$. Then we say that the collection is an *open cover* of A. A *subcover* is a subcollection of the open cover which is also an open cover of A. If any open cover of A contains a finite subcover, then we say that A is *compact*.

Proposition 1.12. *If $A \subset E$ is compact, then A has the Bolzano–Weierstrass property.*

Proof. Let $A \subset E$ be compact and $B \subset A$ infinite. Suppose that B has no limit point in A. Then each point $a \in A$ is the centre of a ball $B(a, r_a)$ containing at most one element of B (the point a, if $a \in B$). These balls form an open cover of A, which has a finite subcover. This implies that B is finite, a contradiction. It follows that A has the Bolzano–Weierstrass property. □

If $A \subset E$, then we define the *diameter* of A by

$$\operatorname{diam}(A) = \sup\{\|x - y\| : x, y \in A\}.$$

Notice that, if $A = \emptyset$, then $\operatorname{diam}(A) = -\infty$. The set A is *bounded* if it lies in some open ball centred on the origin. For a nonempty set, being bounded is equivalent to having a finite diameter.

Example. The diameter of an open ball is twice its radius.

Consider an open cover $\{O_i\}_{i \in I}$ of a set A. If there is a number $\alpha > 0$ such that any subset of A with diameter less than α lies in some member O_i of the open cover, then we say that α is a *Lebesgue number* for the cover. We say that A is *sequentially compact* if every sequence contained in A has a convergent subsequence whose limit lies in A. It turns out that these apparently unrelated notions are in fact related.

Lemma 1.1. *If $A \subset E$ is sequentially compact, then any open cover of A has a Lebesgue number.*

Proof. Suppose that A is sequentially compact and let $\{O_i\}_{i \in I}$ be an open cover of A. If $\{O_i\}_{i \in I}$ has no Lebesgue number, then we can find a sequence of subsets (A_n) of A, with $\operatorname{diam}(A_n) < \frac{1}{n}$, each included in no O_i. Let (x_n) be a sequence in A such that $x_n \in A_n$. By hypothesis, (x_n) has a subsequence converging to a point $x \in A$. x belongs to some O_{i_0} and, as O_{i_0} is open, there is an open ball $B(x,r) \subset O_{i_0}$. Let us now take n_0 such that $x_{n_0} \in B(x, \frac{r}{2})$ and $\frac{1}{n_0} < \frac{r}{2}$. If $y \in A_{n_0}$, then

$$\|y - x\| \le \|y - x_{n_0}\| + \|x_{n_0} - x\| < \frac{r}{2} + \frac{r}{2} = r,$$

and so $A_{n_0} \subset B(x,r) \subset O_{i_0}$, a contradiction. The result now follows. $\square$

If E and F are normed vector spaces, A a subset of E and f a mapping from A into F, then we say that f is *uniformly continuous* if the following condition is satisfied:

for all $\epsilon > 0$, there exists $\alpha > 0$ such that if $x, y \in A$ and $\|x - y\|_E < \alpha$,

then $\|f(x) - f(y)\|_F < \epsilon$.

Clearly, if f is uniformly continuous, then f is continuous. It is easy to find examples of continuous mappings which are not uniformly continuous. For example, the function $f : \mathbb{R}_+^* \longrightarrow \mathbb{R}, t \longmapsto \frac{1}{t}$ is continuous, but not uniformly continuous. However, we do have the following result:

Proposition 1.13. *If E and F are normed vector spaces, $A \subset E$ sequentially compact and $f : A \longrightarrow F$ continuous on A, then f is uniformly continuous.*

Proof. Let $\epsilon > 0$. Suppose that f is continuous on A; then for each $a \in A$, there is an open ball $B(a, r_a)$ such that $\|f(x) - f(a)\|_F < \frac{\epsilon}{2}$ if $x \in B(a, r_a) \cap A$. The balls $B(a, r_a)$ form an open cover of A. If A is sequentially compact, then this cover has a Lebesgue number α. If $x, y \in A$ and $\|x - y\|_E < \alpha$, then x and y lie in some ball $B(a, r_a)$ and so

$$\|f(x) - f(y)\|_F \le \|f(x) - f(a)\|_F + \|f(a) - f(y)\|_F < \frac{\epsilon}{2} + \frac{\epsilon}{2} = \epsilon.$$

This proves the result. $\square$

A subset A of E is *totally bounded* if for any $\epsilon > 0$ there is a finite collection of open balls of radius ϵ forming an open cover of A.

Exercise 1.14. Show that a totally bounded set is bounded.

Lemma 1.2. *If $A \subset E$ is sequentially compact, then A is totally bounded.*

Proof. Let A be sequentially compact and suppose that A is not totally bounded. Then there is an $\epsilon > 0$ and an infinite sequence $(a_n) \subset A$ such that

$$a_{n+1} \in A \setminus \{\cup_{i=1}^n B(a_i, \epsilon)\}.$$

Clearly $\|a_i - a_j\| \geq \epsilon$ if $i \neq j$, which implies that the sequence (a_n) has no convergent subsequence. We have thus obtained a contradiction. It follows that A is totally bounded.   □

We are now in a position to show that compactness and sequential compactness are equivalent. In fact, we will prove a little more.

Theorem 1.2. *If E is a normed vector space and $A \subset E$, then the following three conditions are equivalent:*

(a) *A is compact;*
(b) *A has the Bolzano–Weierstrass property;*
(c) *A is sequentially compact.*

Proof. We have already established that (a) $\Longrightarrow$ (b) (Proposition 1.12).

(b) $\Longrightarrow$ (c) Let (x_n) be a sequence in A. If (x_n) does not contain an infinite subset, then (x_n) contains a constant subsequence, which clearly converges to an element of A. On the other hand, if (x_n) contains an infinite subset, then by hypothesis (x_n) has a limit point $x \in A$. It follows that (x_n) has a subsequence converging to x.

(c) $\Longrightarrow$ (a) Let $\{O_i\}_{i \in I}$ be an open cover of A. By Lemma 1.1 this cover has a Lebesgue number α. We now set $\epsilon = \frac{\alpha}{3}$. From Lemma 1.2 we can find $a_1, \ldots, a_s \in A$ such that $A \subset \cup_{k=1}^s B(a_k, \epsilon)$. However, for each k

$$\operatorname{diam} B(a_k, \epsilon) = 2\epsilon = \frac{2\alpha}{3} < \alpha$$

and hence each ball $B(a_k, \epsilon)$ lies in some O_{i_k}. We thus obtain

$$A \subset \cup_{k=1}^s B(a_k, \epsilon) \subset \cup_{k=1}^s O_{i_k},$$

i.e., $\{O_{i_k}\}_{k=1}^s$ is a finite subcover of A.   □

We will see a little further on that in finite-dimensional spaces we can characterize the compact subsets in another way. These are precisely the subsets which are closed and bounded. Before proving this we will establish some other results.

Exercise 1.15. Let E be a vector space, $A \subset E$ and $\|\cdot\|$ and $\|\cdot\|^\times$ equivalent norms on E. Show that A is compact in the normed vector space $(E, \|\cdot\|)$ if and only if A is compact in the normed vector space $(E, \|\cdot\|^\times)$.

Exercise 1.16. Show that a closed subset of a compact set is compact.

Exercise 1.17. Let E and F be normed vector spaces, A a compact subset of E and $f : A \longrightarrow F$ continuous. Show that $f(A)$ is a compact subset of F.

Proposition 1.14. *If E is a normed vector space and $A \subset E$ is compact, then A is closed and bounded.*

Proof. Let $A \subset E$ be compact. If (x_n) is a convergent subsequence of A, then (x_n) has a subsequence converging to an element $x \in A$. However, x must be the limit of (x_n). Thus a convergent sequence of A has its limit in A and so A is closed.

If A is not bounded, then we can construct a sequence $(x_n) \subset A$ such that $\|x_n\| > n$. This sequence cannot have a convergent subsequence, because convergent sequences are bounded. It follows that A is bounded. $\square$

Exercise 1.18. Show that a nonempty compact subset of a normed vector space has a boundary point.

It is difficult in general to know whether a real-valued function attains a maximum or a minimum on a given set. However, for compact sets we have the following result:

Theorem 1.3. *If E is a normed vector space, $A \subset E$ compact and $f : A \longrightarrow \mathbb{R}$ continuous, then f is bounded on A and attains its lower and upper bounds.*

Proof. If $f(A)$ is not bounded, then there is a sequence $(x_n) \subset A$ such that $|f(x_n)| > n$. As A is sequentially compact, (x_n) has a subsequence (y_n) converging to a point $x \in A$. However, f is continuous and so $\lim f(y_n) = f(x)$ and hence the sequence $(f(y_n))$ is bounded, which is a contradiction. It follows that $f(A)$ is bounded.

Let (x_n) be a sequence in A such that

$$f(x_{n+1}) \leq f(x_n) \qquad \text{and} \qquad \lim f(x_n) = \inf_{x \in A} f(x).$$

The sequence (x_n) has a convergent subsequence (y_n) whose limit y belongs to A. We have

$$\inf_{x \in A} f(x) = \lim f(x_n) = \lim f(y_n) = f(y).$$

In the same way we can show that there is an element $z \in A$ such that $f(z) = \sup_{x \in A} f(x)$. $\square$

We now turn our attention to the particular case of compactness in the normed vector space $(\mathbb{R}^n, \|.\|_\infty)$.

Lemma 1.3. *A subset A of $(\mathbb{R}^n, \|.\|_\infty)$ which is closed and bounded is compact.*

Proof. First, let us consider the case where $n = 1$. As A is bounded, there is a closed interval $[a, b]$ such that $A \subset [a, b]$. Let (x_k) be a sequence in A and for each $k \in \mathbb{N}$ let us set $X_k = \{x_p : p \geq k\}$ and $a_k = \inf X_k$. The sequence (a_k) is increasing and bounded above (by b) and so has a limit x. For each k we take $y_k \in X_k$ such that $a_k \leq y_k < a_k + \frac{1}{k}$. Then

$$|y_k - x| \leq |y_k - a_k| + |a_k - x| < \frac{1}{k} + |a_k - x|.$$

Therefore $\lim y_k = x$ and, as A is closed, $x \in A$. We have established that A is sequentially compact and so compact.

Now let us consider the case where $n \geq 2$. A is contained in a closed rectangle $[a_1, b_1] \times \cdots \times [a_n, b_n]$. Let (x_k) be a sequence in A. Using superscripts for the coordinates of elements of the sequence, we have $x_k = (x_k^1, \ldots, x_k^n)$. The sequence (x_k^1) contains a subsequence (x_{1k}^1) converging to an element $\lambda^1 \in [a_1, b_1]$, because $[a_1, b_1]$ is compact. We now consider the second coordinate. The sequence (x_{1k}^2) contains a subsequence (x_{2k}^2) converging to an element $\lambda^2 \in [a_2, b_2]$, because $[a_2, b_2]$ is compact. Notice that (x_{2k}^1), being a subsequence of (x_{1k}^1), converges to λ^1. We now take a convergent subsequence (x_{3k}^3) of the sequence (x_{2k}^3). Continuing in the same way we obtain a subsequence (y_k) of (x_k) such that (y_k^s) converges to $\lambda^s \in [a_s, b_s]$ for $s = 1, \ldots, n$. Setting $\lambda = (\lambda^1, \ldots, \lambda^n)$ we have $\lim \|y_k - \lambda\|_\infty = 0$. As A is closed, $\lambda \in A$. Therefore A is sequentially compact and so compact. $\qquad \square$

At the beginning of this chapter we showed that certain norms on $\mathbb{R}^n$ were equivalent. We will now show that all norms on $\mathbb{R}^n$ are equivalent and, as a corollary, that all norms on a finite-dimensional normed vector space are equivalent.

Theorem 1.4. *If $\| \cdot \|$ is a norm defined on $\mathbb{R}^n$, then $\| \cdot \|$ is equivalent to the norm $\| \cdot \|_\infty$. It follows that all norms on $\mathbb{R}^n$ are equivalent.*

Proof. Let (e_i) be the standard basis of $\mathbb{R}^n$. If $x = \sum_{i=1}^n x_i e_i$, then

$$\|x\| \leq \sum_{i=1}^n |x_i| \|e_i\| \leq \left(\sum_{i=1}^n \|e_i\| \right) \|x\|_\infty = \beta \|x\|_\infty,$$

therefore $\| \cdot \|$ is continuous on $(\mathbb{R}^n, \| \cdot \|_\infty)$. Let S be the unit sphere in this space: $S = \{x \in \mathbb{R}^n : \|x\|_\infty = 1\}$. S is closed and bounded and so compact. The function $f : S \longrightarrow \mathbb{R}, x \longmapsto \frac{1}{\|x\|}$ is continuous on S and so by Theorem 1.3 there is a constant $K > 0$ such that $\frac{1}{\|x\|} \leq K$, or $\|x\| \geq \frac{1}{K} = \alpha$. For $x \neq 0$ we have

$$\left\| \frac{x}{\|x\|_\infty} \right\| \geq \alpha \implies \|x\| \geq \alpha \|x\|_\infty,$$

an inequality which is also true for $x = 0$. Therefore

$$\alpha \|x\|_\infty \leq \|x\| \leq \beta \|x\|_\infty$$

for all $x \in \mathbb{R}^n$, which establishes the equivalence of the two norms. $\qquad \square$

Corollary 1.3. *All norms on a finite-dimensional vector space E are equivalent.*

Proof. Let (u_i) be a basis of E. For $x = \sum_{i=1}^n x_i u_i$ we set $\phi(x) = (x_1, \ldots, x_n)$. ϕ is a linear isomorphism from E onto $\mathbb{R}^n$. If $\| \cdot \|_E$ and $\| \cdot \|_E^\times$ are norms on E, then we define norms $\| \cdot \|$ and $\| \cdot \|^\times$ on $\mathbb{R}^n$ by

$$\|y\| = \|\phi^{-1}(y)\|_E \qquad \text{and} \qquad \|y\|^\times = \|\phi^{-1}(y)\|_E^\times.$$

As $\| \cdot \|$ and $\| \cdot \|^\times$ are equivalent, so are $\| \cdot \|_E$ and $\| \cdot \|_E^\times$. $\qquad \square$

We are now in a position to prove the characterization of compact sets to which we referred above.

Theorem 1.5. *The compact subsets of a finite-dimensional normed vector space are the subsets which are closed and bounded.*

Proof. We have already seen that a compact set is closed and bounded, so we only need to prove the converse. Let A be a closed and bounded subset of an n-dimensional normed vector space E. We suppose that ϕ and the norm $\| \cdot \|$ on $\mathbb{R}^n$ are defined as above. The set $\phi(A)$ is closed and bounded with respect to the norm $\| \cdot \|$ and therefore compact, because $\| \cdot \|$ is equivalent to $\| \cdot \|_\infty$. As ϕ^{-1} is continuous, A is compact (Exercise 1.17). $\qquad\qquad\qquad\qquad\qquad\qquad\qquad\qquad\qquad\qquad\qquad\qquad\qquad\quad$ ⊔

Exercise 1.19. Show that in a finite-dimensional normed vector space a bounded subset is totally bounded.

Exercise 1.20. Let K be a closed subset of a finite-dimensional normed vector space E and f a real-valued continuous function defined on K such that

$$\lim_{n \to \infty} \|x_n\| = \infty \Longrightarrow \lim f(x_n) = \infty.$$

Show that f has a minimum on K.

Exercise 1.21. Let A be a noncompact subset of a finite-dimensional normed vector space E. Show that there is a continuous real-valued function defined on A which is not bounded.

Having shown that closed bounded sets in a finite-dimensional normed vector space are compact, it is natural to consider such sets in an infinite-dimensional normed vector space. We will take up this question in the next section.

1.5 Banach Spaces

The notion of a Cauchy sequence in $\mathbb{R}$ can be generalized to normed vector spaces. We say that a sequence (x_k) in a normed vector space E is a *Cauchy sequence* if it satisfies the following property:

for all $\epsilon > 0$, there is an $N(\epsilon) \in \mathbb{N}$ such that $\|u_m - u_n\| < \epsilon$, if $m, n \geq N(\epsilon)$.

It is easy to see that a convergent sequence is a Cauchy sequence and that a Cauchy sequence is bounded. We say that a normed vector space E is *complete*, or a *Banach space*, if every Cauchy sequence in E converges.

Theorem 1.6. *The normed vector space $(\mathbb{R}^n, \| \cdot \|_\infty)$ is a Banach space.*

Proof. Let (x_k) be a Cauchy sequence in $\mathbb{R}^n$. Using superscripts for coordinates of elements of the sequence, we have $x_k = (x_k^1, \ldots, x_k^n)$. For $i = 1, \ldots, n$, the sequence (x_k^i) is a Cauchy sequence. As Cauchy sequences in $\mathbb{R}$ converge, for each

i there is an x^i such that $\lim_{k\to\infty} x_k^i = x^i$. If we set $x = (x^1, \ldots, x^n)$, then it is easy to see that $\lim_{k\to\infty} x_k = x$.   □

Corollary 1.4. *A finite-dimensional normed vector space is a Banach space. In particular, the normed vector spaces* $(\mathbb{R}^n, \|\cdot\|_p)$, *for* $1 \le p < \infty$, *are Banach spaces.*

Proof. Let (u_i) be a basis of the n-dimensional normed vector space $(E, \|\cdot\|)$. If $x = \sum_{i=1}^n x_i u_i$ and we set $\phi(x) = (x_1, \ldots, x_n)$, then ϕ is a linear isomorphism of E onto $\mathbb{R}^n$. Now setting

$$\|\phi^{-1}(y)\|^{\times} = \|y\|_\infty$$

for $y \in \mathbb{R}^n$ we obtain a norm on E. As $(\mathbb{R}^n, \|\cdot\|_\infty)$ is complete, so is $(E, \|\cdot\|^{\times})$. The equivalence of norms on E implies that $(E, \|\cdot\|)$ is also complete.   □

Corollary 1.5. *A finite-dimensional subspace of a normed vector space is closed.*

Proof. If F is a finite-dimensional subspace of a normed vector space $(E, \|\cdot\|)$, then $(F, \|\cdot\|)$ is a Banach space. Let (x_n) be a sequence in F with limit $x \in E$. As (x_n) is convergent, (x_n) is a Cauchy sequence in F and so has a limit $x' \in F$. As the limit of a convergent sequence is unique, $x = x'$ and so $x \in F$. It follows that F is closed.   □

We now return to the question of closed bounded sets in an infinite-dimensional normed vector space. Let E be an infinite-dimensional normed vector space and S its unit sphere:

$$S = \{x \in E : \|x\| = 1\}.$$

We will construct by induction a sequence $(u_n) \subset S$ such that $\|u_i - u_j\| > \frac{1}{2}$ if $i \ne j$. For u_1 we choose any element of S. Suppose now that we have constructed the first n elements of the sequence and let $F_n = \text{Vect}(u_1, \ldots, u_n)$, i.e., the vector subspace generated by the vectors $u_1, \ldots, u_n$. As F_n is finite-dimensional, S is not included in F_n and so there is an element $x \in S$, which is not in F_n. Also, F_n is closed and so

$$\alpha = \inf_{y \in F_n} \|x - y\| > 0.$$

We now take $\bar{y} \in F_n$ such that $\alpha \le \|x - \bar{y}\| < 2\alpha$ and set $u_{n+1} = \frac{x - \bar{y}}{\|x - \bar{y}\|}$. Clearly $u_{n+1} \in S$. For $k = 1, \ldots, n$, we set $z_k = \bar{y} + \|x - \bar{y}\|u_k$. Then $z_k \in F_n$ and so

$$\|u_{n+1} - u_k\| = \frac{\|x - z_k\|}{\|x - \bar{y}\|} \ge \frac{\alpha}{\|x - \bar{y}\|} > \frac{1}{2}.$$

Therefore we have constructed a sequence (u_n) with the required properties. The existence of such a sequence implies that S, a closed bounded set, cannot be sequentially compact. Hence a closed bounded subset of a normed vector space is not necessarily compact. (Of course such a subset may be compact, for example, a subset composed of a finite number of points.) In fact, we have also proved the following theorem, referred to as Riesz's theorem.

Theorem 1.7. *The unit sphere* S *of a normed vector space* E *is compact if and only if* E *is finite-dimensional.*

Exercise 1.22. Let E be an infinite-dimensional normed vector space and A a subset of E such that int $A \neq \emptyset$. Show that A is not compact.

Remark. The sequence (u_n) constructed above shows that in an infinite-dimensional normed vector space a subset may be bounded without being totally bounded: any open ball of radius $\frac{1}{4}$ can contain at most one element of the sequence (u_n); hence a finite number of such balls cannot cover S.

Let l^∞ be the collection of bounded sequences of real numbers. With the usual operations on sequences:

$$(x_k) + (y_k) = (x_k + y_k) \qquad \text{and} \qquad \lambda(x_k) = (\lambda x_k)$$

for $\lambda \in \mathbb{R}$, l^∞ is a vector space. If we set

$$\|(x_k)\|_\infty = \sup |x_k|,$$

then $\|\cdot\|_\infty$ defines a norm on l^∞. We claim that l^∞ with this norm is a Banach space. To simplify the notation, let us write $\|\cdot\|$ for $\|\cdot\|_\infty$. Let (x_k) be a Cauchy sequence in l^∞. As Cauchy sequences are bounded, there exists $M > 0$ such that $\|x_k\| \leq M$ for all k. If we write x_k^i for the ith coordinate of x_k, i.e., $x_k = (x_k^1, x_k^2, \ldots)$, then the sequence (x_k^i) is a Cauchy sequence and so has a limit x^i. As the Cauchy sequence (x_k^i) is bounded by M, so is the limit x^i. If we now set $x = (x^1, x^2, \ldots)$, then $x \in l^\infty$. We now show that $\lim x_k = x$. Let $\epsilon > 0$ and N be such that $\|x_k - x_l\| < \epsilon$ for $k, l \geq N$. Then for every i and $k, l \geq N$ we have

$$|x_k^i - x^i| \leq |x_k^i - x_l^i| + |x_l^i - x^i| < \epsilon + |x_l^i - x^i|.$$

However, $\lim_{l \to \infty} |x_l^i - x^i| = 0$ and so, for $k \geq N$, we obtain $|x_k^i - x^i| \leq \epsilon$. It follows that $\|x_k - x\| \leq \epsilon$. We have shown that $\lim x_k = x$ and so that $(l^\infty, \|\cdot\|_\infty)$ is a Banach space.

Now suppose that $1 \leq p < \infty$ and let l^p be the collection of sequences (x_n) of real numbers such that $\sum_{n=1}^\infty |x_n|^p < \infty$. If (x_n) and (y_n) are two sequences in l^p, then, using the first section of this chapter, we have for $m \in \mathbb{N}^*$

$$\left(\sum_{n=1}^m |x_n + y_n|^p \right)^{\frac{1}{p}} \leq \left(\sum_{n=1}^m |x_n|^p \right)^{\frac{1}{p}} + \left(\sum_{n=1}^m |y_n|^p \right)^{\frac{1}{p}} \leq \left(\sum_{n=1}^\infty |x_n|^p \right)^{\frac{1}{p}} + \left(\sum_{n=1}^\infty |y_n|^p \right)^{\frac{1}{p}}$$

and it follows that $(x_n + y_n) \in l^p$ and

$$\left(\sum_{n=1}^\infty |x_n + y_n|^p \right)^{\frac{1}{p}} \leq \left(\sum_{n=1}^\infty |x_n|^p \right)^{\frac{1}{p}} + \left(\sum_{n=1}^\infty |y_n|^p \right)^{\frac{1}{p}}.$$

With the usual operations on sequences, l^p is a vector space and, if we set

$$\|(x_n)\|_p = \left(\sum_{n=1}^{\infty} |x_n|^p \right)^{\frac{1}{p}},$$

then $\| \cdot \|_p$ defines a norm on l^p. We leave it to the reader to show that $(l^p, \| \cdot \|_p)$ is a Banach space.

Exercise 1.23. Let A be a nonempty subset of a normed vector space E and C(A) the set of bounded continuous real-valued functions defined on A. With the usual operations

$$(f + g)(x) = f(x) + g(x) \qquad \text{and} \qquad (\lambda f)(x) = \lambda f(x)$$

for $\lambda \in \mathbb{R}$, $C(A)$ is a vector space. Show that a norm may be defined on $C(A)$ by setting

$$\|f\| = \sup_{x \in A} |f(x)|$$

and that $C(A)$ with this norm is a Banach space.

Remark. We may generalize this exercise by taking A to be any metric space. If A is compact, then the bounded continuous real-valued functions defined on A are just the continuous real-valued functions defined on A.

1.6   Linear and Polynomial Mappings

As all norms on $\mathbb{R}^n$ are equivalent, the continuity of a mapping from $\mathbb{R}^n$ into a normed vector space E or of a mapping from a normed vector space E into $\mathbb{R}^n$ is unaffected by the norm on $\mathbb{R}^n$ we choose. In general, we will work with the norm $\| \cdot \|_\infty$.

Consider a linear mapping f from $\mathbb{R}^n$ into a normed vector space E. If (e_i) is the standard basis of $\mathbb{R}^n$ and $x = \sum_{i=1}^{n} x_i e_i$, then

$$\|f(x)\|_E \leq \sum_{i=1}^{n} \|x_i f(e_i)\|_E = \sum_{i=1}^{n} |x_i| \|f(e_i)\|_E \leq M \|x\|_\infty,$$

where $M = \sum_{i=1}^{n} \|f(e_i)\|_E$. Therefore

$$\|f(x) - f(y)\|_E = \|f(x - y)\|_E \leq M \|x - y\|_\infty$$

and so f is continuous. More generally, a linear mapping from a finite-dimensional normed vector space into another normed vector space is always continuous.

We now consider the continuity of polynomials (in several variables). A constant function $f : \mathbb{R}^n \longrightarrow \mathbb{R}$ is clearly continuous. As the projection $p_i : \mathbb{R}^n \longrightarrow \mathbb{R}, (x_1, \ldots, x_n) \longmapsto x_i$ is linear, p_i is continuous. Also, a polynomial is a sum of products of constant mappings and projections and so continuous.

Let E and $F_1, \ldots, F_p$ be normed vector spaces, $A \subset E$ and f a mapping from A into $F_1 \times \cdots \times F_p$. The vector $f(x)$ has p coordinates which we may write $f_1(x), \ldots, f_p(x)$. We thus obtain p mappings $f_i : A \longrightarrow F_i$. We call these mappings the *coordinate mappings* (or *functions*) of f. A particular case is when $F_i = \mathbb{R}$ for all i; in this case $F_1 \times \cdots \times F_p = \mathbb{R}^p$.

Proposition 1.15. *The mapping $f : A \longrightarrow F_1 \times \cdots \times F_p$ is continuous at $a \in A$ if and only if its coordinate mappings are continuous at a.*

Proof. Suppose first that f is continuous at $a \in A$ and let us take $\epsilon > 0$. There is a $\delta > 0$ such that

$$x \in A \text{ and } \|x - a\|_E < \delta \Longrightarrow \|f(x) - f(a)\|_M < \epsilon,$$

where $\| \cdot \|_M$ is the usual norm on $F_1 \times \cdots \times F_p$. This implies that, for $i = 1, \ldots, p$,

$$\|f_i(x) - f_i(a)\|_{F_i} < \epsilon$$

and it follows that the f_i are continuous at a.

Conversely, suppose that the f_i are continuous at a and let us take $\epsilon > 0$. For each i, there is a $\delta_i > 0$ such that

$$x \in A \text{ and } \|x - a\|_E < \delta_i \Longrightarrow \|f_i(x) - f_i(a)\|_{F_i} < \epsilon.$$

If we set $\delta = \min \delta_i$, then for $i = 1, \ldots, p$

$$x \in A \text{ and } \|x - a\|_E < \delta \Longrightarrow \|f_i(x) - f_i(a)\|_{F_i} < \epsilon$$
$$\Longrightarrow \|f(x) - f(a)\|_M < \epsilon$$

and it follows that f is continuous at a. $\square$

Example. From the above, if $f : \mathbb{R}^n \longrightarrow \mathbb{R}^m$ is such that each coordinate function is a polynomial, then f is continuous.

Let us return to linear mappings. We have seen that a linear mapping from a finite-dimensional vector space into a normed vector space is continuous. However, this is not in general the case. Here is an example. We may define a norm $\| \cdot \|$ on the vector space of polynomials in one variable in the following way: for $p(x) = \sum_{i=1}^n a_i x^i$ we set $\|p\| = \max |a_i|$. The mapping

$$\phi : E \longrightarrow E, p \longmapsto \dot{p},$$

where $\dot{p}$ is the derivative of p, is clearly linear. Consider the sequence of polynomials (p_n) defined by $p_n(x) = \frac{1}{n}x^n$. Clearly $\lim p_n = 0$. However, $\|\dot{p}_n\| = 1$ for all n and so ϕ is not continuous. The next result characterizes continuous linear mappings.

Theorem 1.8. *Let E and F be normed vector spaces and ϕ a linear mapping from E into F. Then the following statements are equivalent:*

(a) *ϕ is continuous;*
(b) *ϕ is continuous at 0;*
(c) *ϕ is bounded on the closed unit ball $\bar{B}(0, 1)$ of E;*
(d) *There exists $\mu \in \mathbb{R}_+$ such that*

$$\|\phi(x)\|_F \leq \mu \|x\|_E$$

for all $x \in E$.

Proof. (a) $\Longrightarrow$ (b) This is true by the definition of continuity.
(b) $\Longrightarrow$ (c) If f is continuous at 0, then there exists $\alpha > 0$ such that

$$\|x - 0\|_E \leq \alpha \Longrightarrow \|\phi(x) - \phi(0)\|_F \leq 1,$$

i.e.,

$$\|x\|_E \leq \alpha \Longrightarrow \|\phi(x)\|_F \leq 1.$$

If $x \in \bar{B}(0, 1)$, then $\|\alpha x\|_E \leq \alpha$ and so $\|\phi(\alpha x)\|_F \leq 1$. Using the linearity of ϕ we obtain $\alpha \|\phi(x)\|_F \leq 1$ and hence the result.
(c) $\Longrightarrow$ (d) By hypothesis, there exists $\mu \in \mathbb{R}_+$ such that

$$x \in \bar{B}(0, 1) \Longrightarrow \|\phi(x)\|_F \leq \mu.$$

If $x \in E \setminus \{0\}$, then $\frac{x}{\|x\|} \in \bar{B}(0, 1)$ and so $\|\phi(\frac{x}{\|x\|})\|_F \leq \mu$. It follows that $\|\phi(x)\|_F \leq \mu \|x\|_E$, which is also true when $x = 0$.
(d) $\Longrightarrow$ (a) If (d) holds and $x, y \in E$, then

$$\|\phi(x) - \phi(y)\|_F = \|\phi(x - y)\|_F \leq \mu \|x - y\|_E,$$

therefore ϕ is continuous. $\qquad\qquad\square$

The continuous linear mappings between two normed vector spaces E and F form a vector space $\mathcal{L}(E, F)$. If we set

$$|\phi|_{\mathcal{L}(E,F)} = \sup_{\|x\|_E \leq 1} \|\phi(x)\|_F$$

for $\phi \in \mathcal{L}(E, F)$, then $|\cdot|_{\mathcal{L}(E,F)}$ is a norm on $\mathcal{L}(E, F)$. Notice that for any $x \in E$ we have

$$\|\phi(x)\|_F \leq |\phi|_{\mathcal{L}(E,F)} \|x\|_E.$$

If G is another normed vector space and $\psi \in \mathcal{L}(F, G)$, then

$$|\psi \circ \phi|_{\mathcal{L}(E,G)} \leq |\psi|_{\mathcal{L}(F,G)} |\phi|_{\mathcal{L}(E,F)}$$

We usually write $\mathcal{L}(E)$ for $\mathcal{L}(E, E)$ and E^* for $\mathcal{L}(E, \mathbb{R})$. E^* is called the *dual* of E and its elements are often referred to as *linear forms*. When there is no confusion possible, we will usually write $|\phi|$ for $|\phi|_{\mathcal{L}(E,F)}$.

Exercise 1.24. Show that

$$|\phi|_{\mathcal{L}(E,F)} = \inf\{\mu \in \mathbb{R}_+ : \|\phi(x)\|_F \leq \mu \|x\|_E, x \in E\}$$

and, if $\dim E \geq 1$, then

$$|\phi|_{\mathcal{L}(E,F)} = \sup_{\|x\|_E = 1} \|\phi(x)\|_F.$$

There is a natural question, namely when is the space $\mathcal{L}(E, F)$ complete, i.e., a Banach space. The following result gives us a sufficient condition.

Theorem 1.9. *If E and F are normed vector spaces and F is complete, then the space $\mathcal{L}(E, F)$ is complete.*

Proof. Let (ϕ_n) be a Cauchy sequence in $\mathcal{L}(E, F)$. Then for each $x \in E$ we have

$$\|\phi_n(x) - \phi_m(x)\|_F \leq |\phi_n - \phi_m|_{\mathcal{L}(E,F)} \|x\|_E$$

and so $(\phi_n(x))$ is a Cauchy sequence in F. As F is complete, this sequence has a limit, which we will write $\phi(x)$. It is easy to see that ϕ is a linear mapping. As (ϕ_n) is a Cauchy sequence, the norms $|\phi_n|$ have an upper bound M. Therefore

$$\|\phi(x)\|_F = \|\lim \phi_n(x)\|_F = \lim \|\phi_n(x)\|_F \leq M \|x\|_E$$

and so ϕ is continuous. It remains to show that $\lim \phi_n = \phi$. Let $\epsilon > 0$ and $N \in \mathbb{N}^*$ be such that $m, n \geq N \Longrightarrow |\phi_m - \phi_n| < \frac{\epsilon}{2}$. If $\|x\|_E \leq 1$ and $m, n \geq N$, then

$$\|\phi_n(x) - \phi_m(x)\|_F \leq |\phi_n - \phi_m| \|x\|_E < \frac{\epsilon}{2}.$$

We may choose $m \geq N$ such that $\|\phi_m(x) - \phi(x)\|_F < \frac{\epsilon}{2}$ and so

$$\|\phi_n(x) - \phi(x)\|_F \leq \|\phi_n(x) - \phi_m(x)\|_F + \|\phi_m(x) - \phi(x)\|_F < \epsilon.$$

As $\|\phi_n(x) - \phi(x)\|_F < \epsilon$ for every x such that $\|x\|_E \leq 1$, we have $|\phi_n - \phi| \leq \epsilon$ and it follows that $\lim \phi_n = \phi$. This ends the proof. $\square$

Let E and $F_1, \ldots, F_p$ be normed vector spaces and let us set $F = F_1 \times \cdots \times F_p$. The mapping f from E into F is linear and continuous if and only if its coordinate mappings are also linear and continuous. Therefore the mapping $\Phi : f \longmapsto (f_1, \ldots, f_p)$, when restricted to $\mathcal{L}(E, F)$, has its image in $\mathcal{L}(E, F_1) \times \cdots \times \mathcal{L}(E, F_p)$. There is no difficulty in seeing that Φ defines a linear isomorphism between the two spaces. Also,

$$
\begin{aligned}
|f|_{\mathcal{L}(E,F)} &= \sup_{\|x\|_E \leq 1} \|f(x)\|_F \\
&= \sup_{\|x\|_E \leq 1} \max(\|f_1(x)\|_{F_1}, \ldots, \|f_p(x)\|_{F_p}) \\
&= \max(\sup_{\|x\|_E \leq 1} \|f_1(x)\|_{F_1}, \ldots, \sup_{\|x\|_E \leq 1} \|f_p(x)\|_{F_p}) \\
&= \|(|f_1|_{\mathcal{L}(E,F_1)}, \ldots, |f_p|_{\mathcal{L}(E,F_p)})\|_M,
\end{aligned}
$$

where $\| \cdot \|_M$ is the usual norm on the cartesian product of the spaces $\mathcal{L}(E, F_i)$. Therefore Φ also preserves the norm.

If E and F are normed vector spaces and $f : E \longrightarrow F$ a continuous linear isomorphism whose inverse f^{-1} is also continuous, then we say that f is a *normed vector space isomorphism* and that the spaces E and F are isomorphic (as normed vector spaces). In this case, we write $E \simeq F$. If $f : E \longrightarrow F$ is a linear isomorphism which preserves the norm, i.e., $\|f(x)\|_F = \|x\|_E$ for all $x \in E$, then we say that f is an *isometric isomorphism*. Clearly an isometric isomorphism is a normed vector space isomorphism. If there exists an isometric isomorphism from E onto F, then we say that the two spaces are isometrically isomorphic.

Exercise 1.25. Show that two finite-dimensional normed vector spaces of the same dimension are isomorphic.

Exercise 1.26. Let E be a normed vector space. Show that the mapping $\Phi : \mathcal{L}(\mathbb{R}, E) \longrightarrow E, f \longmapsto f(1)$ is an isometric isomorphism and thus that $\mathcal{L}(\mathbb{R}, E)$ and E are isomorphic.

Exercise 1.27. Show that if E, F, and G are normed vector spaces and $\phi : F \longrightarrow G$ is a normed vector space isomorphism, then the mapping Φ from $\mathcal{L}(E, F)$ into $\mathcal{L}(E, G)$, defined by $\Phi(f) = \phi \circ f$, is a normed vector space isomorphism. Show that Φ is an isometric isomorphism if ϕ is an isometric isomorphism.

Suppose now that $\phi : E \longrightarrow F$ is a normed vector space isomorphism. Clearly, if the dimensions of E and F are 0, then ϕ is norm-preserving. Suppose now that this is not the case. We can define a norm $\| \cdot \|_E^\times$ on E by setting

$$
\|x\|_E^\times = \|\phi(x)\|_F.
$$

We have

$$
\|x\|_E^\times = \|\phi(x)\|_F \leq |\phi| \|x\|_E
$$

and
$$\|x\|_E = \|\phi^{-1}(\phi(x))\|_E \leq |\phi^{-1}|\|\phi(x)\|_F = |\phi^{-1}|\|x\|_E^\times.$$
Because $|\phi^{-1}| \neq 0$, we can write
$$\frac{1}{|\phi^{-1}|}\|x\|_E \leq \|x\|_E^\times \leq |\phi|\|x\|_E.$$

We have shown that we can give E an equivalent norm such that f is an isometric isomorphism.

1.7   Normed Algebras

A vector space E may have a multiplication as well as its addition and scalar multiplication. In this case, if E with its addition and multiplication is a ring and satisfies the property:
$$\lambda(xy) = (\lambda x)y = x(\lambda y)$$
for $\lambda \in \mathbb{R}$ and $x, y \in E$, then we say that E is an *algebra*. If the multiplication is commutative, then we say that the algebra is a *commutative algebra*. We usually assume that there is an identity for the multiplication, for which we use the symbol 1_E (or just 1). If E is an algebra and has a norm $\|\cdot\|$ satisfying the properties
$$\|xy\| \leq \|x\|\|y\| \qquad \text{and} \qquad \|1_E\| = 1,$$

then we say that E is a *normed algebra*. If E is also complete, then we say that E is a *Banach algebra*. If E and F are normed algebras and $f \in \mathcal{L}(E, F)$ such that f is also a ring homomorphism, then we say that f is a *normed algebra homomorphism*. If f is a normed vector space isomorphism, then f is said to be a normed algebra isomorphism and E and F isomorphic normed algebras.

Exercise 1.28. Show that if E is a normed algebra, then the multiplication as a mapping from E^2 into E is continuous.

Let us look at some examples. $\mathbb{R}$ has a natural vector space structure and, with the usual multiplication, is a commutative algebra. The absolute value defines a norm on $\mathbb{R}$ and, with this norm, $\mathbb{R}$ is a normed algebra. As $\mathbb{R}$ is complete, $\mathbb{R}$ is a commutative Banach algebra.

$\mathbb{R}^2$ also has a natural vector space structure and, with the multiplication

$$(x, y) \cdot (u, v) = (xu - yv, xv + yu),$$

$\mathbb{R}^2$ is a commutative algebra. If we give $\mathbb{R}^2$ the norm $\|\cdot\|_2$, $\mathbb{R}^2$ becomes a normed algebra. As $\mathbb{R}^2$ is complete, $\mathbb{R}^2$ is a commutative Banach algebra. (Of course, the multiplication we have here is that used in defining the field of complex numbers.)

We have already seen that, if E is a normed vector space, then we can define a norm on $\mathcal{L}(E)$ by setting

$$|\phi|_{\mathcal{L}(E)} = \sup_{\|x\| \leq 1} \|\phi(x)\|.$$

With this norm $\mathcal{L}(E)$ is a normed algebra. From Theorem 1.9, if E is a Banach space, then $\mathcal{L}(E)$ is a Banach algebra.

Here is another example of a normed algebra. Consider the set $\mathbb{R}[X]$ of real polynomials in one variable. $\mathbb{R}[X]$ is clearly an algebra and has the polynomial $P \equiv 1$ for multiplicative identity. If, for $P \in \mathbb{R}[X]$ we set

$$\|P\| = \sup_{x \in [0,1]} |P(x)|,$$

then we obtain a normed algebra.

In Exercise 1.23 we introduced the Banach space $C(A)$ of bounded continuous real-valued functions defined on a metric space, in particular, on a nonempty subset of a normed vector space. If we now add the multiplication defined by

$$(f \cdot g)(x) = f(x)g(x),$$

then $C(A)$ becomes a Banach algebra.

Notation. We will write $\mathcal{M}_{mn}(\mathbb{R})$ for the set of $m \times n$ real matrices; if $m = n$, then we will use the shorter notation $\mathcal{M}_n(\mathbb{R})$ instead of $\mathcal{M}_{nn}(\mathbb{R})$.

As a final example, let us consider the space $\mathcal{M}_n(\mathbb{R})$, with $n \geq 1$. There is no difficulty in seeing that $\mathcal{M}_n(\mathbb{R})$ with the usual operations on matrices is an algebra. Suppose now that $\| \cdot \|$ is a norm defined on $\mathbb{R}^n$. If, for $A \in \mathcal{M}_n(\mathbb{R})$ we set

$$|A| = \sup_{\|x\|=1} \|Ax\|,$$

then we obtain a normed algebra. As $\mathcal{M}_n(\mathbb{R})$ is finite-dimensional, $\mathcal{M}_n(\mathbb{R})$ is a Banach algebra. We say that the norm we have just defined is *subordinate* to the norm $\| \cdot \|$.

An element x in a normed algebra E is *invertible* (or *regular*), if there exists $y \in E$ such that $xy = yx = 1$. Otherwise we say that x is *noninvertible* (or *singular*). If x is invertible, then the y is unique and we call this element the inverse of x and write x^{-1} for it. Clearly 0 is not invertible; however, other elements may also be noninvertible. We will write $E^{\times}$ for the set of invertible elements of E. $E^{\times}$, with the multiplication of E, is a group.

Exercise 1.29. Let E be a normed algebra. Show that $\|x^n\| \leq \|x\|^n$ for all $x \in E$ and $n \in \mathbb{N}$. What can we say if x is invertible and $n \in \mathbb{Z}_- = \{n \in \mathbb{Z} : n < 0\}$?

We aim to look at the mapping $\phi : x \longmapsto x^{-1}$ in some detail; however, we will need a preliminary result.

Lemma 1.4. *If E is a Banach algebra and $\|1 - x\| < 1$, then x is invertible.*

Proof. Let $r = \|1 - x\|$. Then for $n \in \mathbb{N}$

$$\|(1 - x)^n\| \leq \|1 - x\|^n = r^n.$$

It follows that the partial sums of the power series $\sum_{n \geq 0}(1 - x)^n$ form a Cauchy sequence in E. As E is complete, the partial sums converge to an element of E, which we will denote $\sum_{n=0}^{\infty}(1 - x)^n$. If we set $y = \sum_{n=0}^{\infty}(1 - x)^n$, then

$$y - xy = (1 - x)y = \sum_{n=1}^{\infty}(1 - x)^n = y - 1,$$

which implies that $xy = 1$. We can show in the same way that $yx = 1$ and so x is invertible and $x^{-1} = \sum_{n=0}^{\infty}(1 - x)^n$. $\qquad\square$

Theorem 1.10. *If E is a Banach algebra, then $E^{\times}$ is open and the mapping $\phi : x \longmapsto x^{-1}$ is continuous and hence a homeomorphism from $E^{\times}$ onto itself.*

Proof. Let $a \in E^{\times}$ and let us set $r = \frac{1}{\|a^{-1}\|}$. If x lies in the open ball $B(a, r)$, then

$$\|1 - a^{-1}x\| = \|a^{-1}(a - x)\| \leq \|a^{-1}\|\|a - x\| < 1,$$

therefore, from the lemma, $a^{-1}x \in E^{\times}$. This implies that $x = a(a^{-1}x) \in E^{\times}$. Thus the ball $B(a, r) \subset E^{\times}$ and it follows that $E^{\times}$ is open. Also, as

$$x^{-1}a = (a^{-1}x)^{-1} = \sum_{n=0}^{\infty}(1 - a^{-1}x)^n,$$

we have

$$\|x^{-1} - a^{-1}\| = \|(x^{-1}a - 1)a^{-1}\| \leq \|a^{-1}\|\|x^{-1}a - 1\| = \|a^{-1}\|\|\sum_{n=1}^{\infty}(1 - a^{-1}x)^n\|.$$

If we take $x \in B(a, \frac{r}{2})$, then $\|1 - a^{-1}x\| < \frac{1}{2}$ and

$$\|x^{-1} - a^{-1}\| \leq \|a^{-1}\|\sum_{n=1}^{\infty}\|(1 - a^{-1}x)\|^n \leq 2\|a^{-1}\|\|1 - a^{-1}x\| \leq 2\|a^{-1}\|^2\|a - x\|,$$

and so ϕ is continuous at a. $\qquad\square$

Let E and F be normed vector spaces. We will write $\mathcal{I}(E, F)$ for the subset of invertible mappings belonging to $\mathcal{L}(E, F)$. Clearly $\mathcal{I}(E, E) = \mathcal{L}(E)^{\times}$. From what we have seen above, if E is a Banach space, then $\mathcal{L}(E)^{\times}$ is an open subset of $\mathcal{L}(E)$ and the mapping $\phi : u \longrightarrow u^{-1}$ is a homeomorphism from $\mathcal{L}(E)^{\times}$ onto itself. We will now generalize this result.

Theorem 1.11. *If E and F are Banach spaces, then $\mathcal{I}(E, F)$ is open in $\mathcal{L}(E, F)$ and $\mathcal{I}(F, E)$ is open in $\mathcal{L}(F, E)$. If $\mathcal{I}(E, F)$ is not empty, then the mapping ψ : $u \longmapsto u^{-1}$ is a homeomorphism from $\mathcal{I}(E, F)$ onto $\mathcal{I}(F, E)$.*

Proof. If $\mathcal{I}(E, F)$ is empty, then it is open. If this is not the case, then let $w \in \mathcal{I}(E, F)$. If we set $\alpha(u) = w^{-1} \circ u$ for $u \in \mathcal{L}(E, F)$, then $\alpha(u) \in \mathcal{L}(E)$ and α is a normed vector space isomorphism from $\mathcal{L}(E, F)$ onto $\mathcal{L}(E)$. As $\alpha(\mathcal{I}(E, F)) = \mathcal{I}(E), \mathcal{I}(E, F)$ is open in $\mathcal{L}(E, F)$. In the same way we may show that $\mathcal{I}(F, E)$ is open in $\mathcal{L}(F, E)$.

The mapping ψ may be written $\psi = \beta \circ \phi \circ \alpha$, where ϕ is the inversion mapping on $\mathcal{L}(E)^{\times}$ and β the normed vector space isomorphism from $\mathcal{L}(E)$ onto $\mathcal{L}(F, E)$ defined by $\beta(v) = v \circ w^{-1}$. As ψ is a composition of three continuous mappings, ψ is continuous. Also, as $\psi^{-1} = \alpha^{-1} \circ \phi \circ \beta^{-1}$, ψ^{-1} is continuous and so ψ is a homeomorphism from $\mathcal{I}(E, F)$ onto $\mathcal{I}(F, E)$. $\qquad\square$

In an algebra certain types of vector subspace deserve particular attention. An *ideal* in an algebra A is a vector subspace I such that, when $a \in A$ and $x \in I$, then

(a) $ax \in I$;
(b) $xa \in I$.

If the first (resp. second) condition is satisfied (and not necessarily both of them), then I is called a *left-sided ideal* (resp. *right-sided ideal*). An ideal I is a *maximal ideal* if there is no ideal $J \neq A$ which properly contains I. Using Zorn's lemma it is easy to show that a proper ideal is always contained in a maximal ideal. We may define maximal left- and right-sided ideals in a similar way and obtain analogous results. As $\{0\}$ is an ideal, an algebra contains all three types of maximal ideal.

The intersection of all left-sided maximal ideals is a left-sided ideal. We have a similar statement for right-sided ideals. It turns out that these intersections are the same. This subset of A is an ideal, which is called the Jacobson radical of A, usually noted $J(A)$. It is also the case that $a \in J(A)$ if and only if $1 - xay$ is invertible for all $x, y \in A$. (Proofs of the statements given here may be found in any standard text on ring theory, for example [9].)

Exercise 1.30. Let A be an algebra and I a left-sided ideal of A as a ring, i.e., I is a subgroup of the group $(A, +)$ satisfying the condition (a). Show that I is a vector subspace of A and thus that I is a left-sided ideal of A as an algebra.

Exercise 1.31. Let E be a normed algebra and I a left-sided ideal in E. Show that $\bar{I}$ is a left-sided ideal and deduce that a maximal left-sided ideal is always closed. It follows that the Jacobson radical is closed.

We will finish this section with an introduction to the *quaternion algebra*. We define a multiplication on $\mathbb{R}^4$ by setting

$$
\begin{aligned}
x \cdot y &= (x_1 y_1 - x_2 y_2 - x_3 y_3 - x_4 y_4)e_1 \\
&= +(x_1 y_2 + x_2 y_1 + x_3 y_4 - x_4 y_3)e_2 \\
&= +(x_1 y_3 - x_2 y_4 + x_3 y_1 + x_4 y_2)e_3 \\
&= +(x_1 y_4 + x_2 y_3 - x_3 y_2 + x_4 y_1)e_4,
\end{aligned}
$$

where $e_1, \ldots, e_4$ is the standard basis of $\mathbb{R}^4$. With this multiplication and its standard addition, $\mathbb{R}^4$ is an algebra with e_1 as multiplicative identity. We usually write $\mathbb{H}$ for this algebra (after William Hamilton who was the first to discover it) and, in this context, $(\mathbf{1}, \mathbf{i}, \mathbf{j}, \mathbf{k})$ for the standard basis. It is easy to check that

$$\mathbf{i}^2 = \mathbf{j}^2 = \mathbf{k}^2 = -\mathbf{1}$$

and

$$\mathbf{ij} = -\mathbf{ji} = \mathbf{k}, \qquad \mathbf{jk} = -\mathbf{kj} = \mathbf{i} \qquad \text{and} \qquad \mathbf{ki} = -\mathbf{ik} = \mathbf{j}.$$

Thus $\mathbb{H}$ is noncommutative. We may also check that, with the norm $\| \cdot \|_2$, $\mathbb{H}$ is a normed algebra. Also, we have

$$\|xy\| = \|x\|\|y\|.$$

If $x \in \mathbb{H}$ and we set

$$x^* = (x_1, -x_2, -x_3, -x_4),$$

then we call x^* the *conjugate of* x and we have

$$(xy)^* = y^* x^* \qquad \text{and} \qquad \|x\|^2 = x^* x = xx^*.$$

From the last property, we see that every nonzero element of $\mathbb{H}$ is invertible: $x^{-1} = \frac{x^*}{\|x\|^2}$, i.e., $\mathbb{H}^\times = \mathbb{H}^*$. An algebra A such that $A^\times = A^*$ is called a *division algebra*. Other examples are $\mathbb{R}$ and $\mathbb{C}$. It turns out that all Banach division algebras are isomorphic to $\mathbb{R}$, $\mathbb{C}$, or $\mathbb{H}$. (A proof of this may be found in [4].)

1.8   The Exponential Mapping

In this section we will see how the exponential mapping of elementary calculus generalizes to Banach algebras. Let E be a Banach algebra and u an element of E. We set

$$s_n = 1 + u + \frac{u^2}{2!} + \cdots + \frac{u^n}{n!}.$$

If $p > q$, then

$$\|s_p - s_q\| = \| \sum_{i=q+1}^{p} \frac{u^i}{i!} \| \le \sum_{i=q+1}^{p} \frac{\|u\|^i}{i!}.$$

Therefore (s_n) is a Cauchy sequence and so converges. We write $\exp(u)$ for the limit of the sequence (s_n). The mapping

$$\exp : E \longrightarrow E, u \longmapsto \exp(u)$$

is called the exponential mapping on E. Notice that $\| \exp(u) \| \le e^{\|u\|}$, where e is the usual exponential mapping on $\mathbb{R}$.

Lemma 1.5. *If E is a normed algebra, u and v elements of E, $M = \max(\|u\|, \|v\|)$ and $n \in \mathbb{N}^*$, then*

$$\|u^n - v^n\| \le nM^{n-1}\|u - v\|.$$

Proof. For $n = 1$ the result is clear. Suppose that $n \ge 2$. Then we can write

$$u^n - v^n = u^{n-1}(u - v) + u^{n-2}(u - v)v + u^{n-3}(u - v)v^2 + \cdots + (u - v)v^{n-1}$$

and the result follows.   □

Proposition 1.16. *The exponential mapping is continuous.*

Proof. Let E be a Banach algebra and $u, v \in E$. If $M = \max(\|u\|, \|v\|)$ and $n \in \mathbb{N}^*$, then from Lemma 1.5 we have

$$\| \frac{u^n}{n!} - \frac{v^n}{n!} \| \le \frac{M^{n-1}}{(n-1)!} \|u - v\|$$

and so

$$\| \exp(u) - \exp(v) \| = \left\| \sum_{n=1}^{\infty} \frac{u^n - v^n}{n!} \right\| \le \sum_{n=1}^{\infty} \frac{M^{n-1}}{(n-1)!} \|u - v\| = e^M \|u - v\|.$$

It follows that the mapping exp is continuous.   □

Exercise 1.32. Generalize the real-valued functions sin, cos, sinh, and cosh to Banach algebras and show that they are continuous.

The exponential mapping e defined on $\mathbb{R}$ has the property: $e^{x+y} = e^x e^y$. This property carries over to Banach algebras for commuting elements.

Theorem 1.12. *If E is a Banach algebra and u and v commuting elements of E, then*

$$\exp(u + v) = \exp(u) \exp(v).$$

Proof. For $m \in \mathbb{N}$ let

$$A_m = \sum_{k=0}^{2m} \frac{(u+v)^k}{k!} - \left(\sum_{i=1}^{m} \frac{u^i}{i!}\right)\left(\sum_{j=1}^{m} \frac{v^j}{j!}\right).$$

Because $uv = vu$, we have

$$\frac{(u+v)^k}{k!} = \frac{1}{k!}\sum_{i=0}^{k} \frac{k!}{i!(k-i)!}u^i v^{k-i} = \sum_{i+j=k} \frac{1}{i!j!}u^i v^j.$$

Therefore

$$A_m = \sum_{0 \le i+j \le 2m} \frac{u^i}{i!}\frac{v^j}{j!} - \sum_{0 \le i \le m, 0 \le j \le m} \frac{u^i}{i!}\frac{v^j}{j!}$$

$$= \sum_{k=0}^{m-1} \frac{u^k}{k!} \sum_{j=m+1}^{2m-k} \frac{v^j}{j!} + \sum_{k=0}^{m-1} \frac{v^k}{k!} \sum_{j=m+1}^{2m-k} \frac{u^j}{j!},$$

and so

$$\|A_m\| \le \sum_{k=0}^{m-1} \frac{\|u\|^k}{k!} \sum_{j=m+1}^{2m-k} \frac{\|v\|^j}{j!} + \sum_{k=0}^{m-1} \frac{\|v\|^k}{k!} \sum_{j=m+1}^{2m-k} \frac{\|u\|^j}{j!}$$

$$= \sum_{k=0}^{2m} \frac{(\|u\|+\|v\|)^k}{k!} - \left(\sum_{i=0}^{m} \frac{\|u\|^i}{i!}\right)\left(\sum_{j=0}^{m} \frac{\|v\|^j}{j!}\right).$$

Letting m go to ∞, we obtain

$$\|\exp(u+v) - \exp(u)\exp(v)\| \le e^{\|u\|+\|v\|} - e^{\|u\|}e^{\|v\|} = 0.$$

This proves the theorem. $\square$

Corollary 1.6. *If E is a Banach algebra, then the image of the exponential mapping is contained in $E^\times$.*

Proof. As x and $-x$ commute, we have

$$\exp(-x)\exp(x) = \exp(0) = 1$$

and so $\exp(x) \in E^\times$. $\square$

Exercise 1.33. Establish the identity

$$\exp(u) = \lim_{n\to\infty} \left(1 + \frac{u}{n}\right)^n.$$

Appendix: The Fundamental Theorem of Algebra

Every nonconstant complex polynomial has a complex root. This is the so-called *fundamental theorem of algebra*. This result can be proved using the existence of a global minimum of a continuous function defined on a compact set. Before beginning the proof, we recall that $\mathbb{C}$, the field of complex numbers, is $\mathbb{R}^2$ equipped with the usual componentwise addition and the multiplication:

$$(x, y).(u, v) = (xu - yv, xv + yu).$$

If $z = (x, y)$, then the modulus of z, written $|z|$, is defined by $|z| = \sqrt{x^2 + y^2}$, i.e., $|z| = \|z\|_2$. It is easy to see that a complex polynomial, i.e., a polynomial mapping with complex coefficients, is a continuous function from $\mathbb{R}^2$ into itself.

Lemma 1.6. *Let P be a nonconstant polynomial of degree n and $z_0 \in \mathbb{C}$ such that $P(z_0) \neq 0$. Then for any $\epsilon > 0$ there exists $z \in \mathbb{C}$ such that $|z - z_0| < \epsilon$ and $|P(z)| < |P(z_0)|$.*

Proof. Let $Q(z) = \frac{1}{P(z_0)} P(z_0 + z)$. Then $\deg Q = n$ and $Q(0) = 1$. Hence we can write

$$Q(z) = 1 + a_p z^p + a_{p+1} z^{p+1} + \cdots + a_n z^n = 1 + a_p z^p + z^p T(z),$$

where $a_p \neq 0$, $p \geq 1$ and $T(0) = 0$. To simplify the notation let us write a for a_p. As T is continuous, there is an $\alpha > 0$ such that $|T(z)| < \frac{|a|}{2}$ for $|z| < \alpha$. For such z we have

$$|Q(z)| \leq |1 + az^p| + \frac{1}{2}|a||z|^p.$$

We now set $\theta = \frac{1}{p}(\pi - \arg a)$, where $\arg a$ is the argument of a in the interval $[0, 2\pi)$. We take $\epsilon > 0$ and choose r such that

$$0 < r < \min\left(\epsilon, \alpha, \frac{1}{|a|^{\frac{1}{p}}}\right)$$

and set $u = re^{i\theta}$. Then

$$au^p = ar^p e^{ip\theta} = ar^p e^{i(\pi - \arg a)} = -|a|r^p,$$

which implies that

$$|au^p| = |a|r^p < 1,$$

because $r < \frac{1}{|a|^{\frac{1}{p}}}$. It follows that

$$1 + au^p = 1 - |a|r^p > 0.$$

Now, $|u| < \alpha$ and so

$$|Q(u)| \le |1 + au^p| + \frac{1}{2}|a||u^p| = 1 - |a|r^p + \frac{1}{2}|a||u^p| = 1 - \frac{1}{2}|a||u^p| < 1.$$

If $z = z_0 + u$, then we have

$$P(z) = P(z_0)Q(u) \implies |P(z)| = |P(z_0)||Q(u)| < |P(z_0)|.$$

In addition,

$$|z - z_0| = |u| = r < \epsilon.$$

This ends the proof.   □

Lemma 1.7. *If P is a nonconstant complex polynomial and $M > 0$, then there is an $r > 0$ such that $|P(z)| > M$ whenever $|z| > r$.*

Proof. If $P(z) = a_0 + a_1 z + \cdots + a_n z^n$, with $a_n \ne 0$, then for $z \ne 0$

$$P(z) = a_n z^n \left[\left(\frac{a_0}{a_n}\right)\frac{1}{z^n} + \left(\frac{a_1}{a_n}\right)\frac{1}{z^{n-1}} + \cdots + \left(\frac{a_{n-1}}{a_n}\right)\frac{1}{z} + 1 \right] = a_n z^n (f(z)+1).$$

Let $A \in \mathbb{R}_+^*$ be such that $A > \max_{0 \le i \le n-1} |\frac{a_i}{a_n}|$. Then

$$|f(z)| \le \left|\frac{a_0}{a_n}\right|\frac{1}{|z|^n} + \cdots + \left|\frac{a_{n-1}}{a_n}\right|\frac{1}{|z|} < A\left(\frac{1}{|z|^n} + \cdots + \frac{1}{|z|}\right).$$

For $|z| \ge \max(1, 2nA)$ we have

$$|f(z)| < An\frac{1}{|z|} \le \frac{1}{2}$$

and so

$$|1 + f(z)| = |1 - (-f(z))| \ge |1 - |f(z)|| = 1 - |f(z)| > 1 - \frac{1}{2} = \frac{1}{2}.$$

This implies that if z is such that $|z| \ge \max(1, 2nA)$, then

$$|P(z)| = |a_n z^n||f(z) + 1| \ge \frac{1}{2}|a_n||z|^n.$$

To conclude, it is sufficient to notice that $\lim_{t \to \infty} |a_n|t^n = \infty$.   □

Having proved the above two lemmas, we are in a position to establish the result stated at the beginning of the appendix.

Theorem 1.13. *A nonconstant complex polynomial P has a root.*

Proof. Suppose that $|P(z)| > 0$ for all $z \in \mathbb{C}$. From Lemma 1.7 there exists $r > 0$ such that $|P(z)| > |P(0)|$ if $|z| > r$. As the closed disc

$$\bar{D}(0, r) = \{z \in \mathbb{C} : |z| \le r\}$$

is compact and P is continuous, there exists $z_0 \in \bar{D}(0, r)$ such that $|P(z)| \ge |P(z_0)|$ when $z \in \bar{D}(0, r)$. If $z \notin \bar{D}(0, r)$, then $|z| > r$ and so $|P(z)| > |P(0)| \ge |P(z_0)|$. Hence $|P(z)| \ge |P(z_0)|$ for all $z \in \mathbb{C}$, which contradicts Lemma 1.6. $\square$

A quite different proof of the fundamental theorem of algebra may be found in [19]. This proof is algebraic in nature, in contrast to the analytical proof given here.

Chapter 2
Differentiation

In this chapter we will be primarily concerned with extending the derivative defined for real-valued functions defined on an interval of $\mathbb{R}$. We will also consider minima and maxima of real-valued functions defined on a normed vector space.

2.1 Directional Derivatives

Let O be an open subset of a normed vector space E, f a real-valued function defined on O, $a \in O$ and u a nonzero element of E. The function $f_u : t \longrightarrow f(a + tu)$ is defined on an open interval containing 0. If the derivative $\frac{df_u}{dt}(0)$ is defined, i.e., if the limit

$$\lim_{t \to 0} \frac{f(a + tu) - f(a)}{t}$$

exists, then we note this derivative $\partial_u f(a)$. It is called the derivative of f at a in the direction u. We refer to such derivatives as *directional derivatives*. Notice that, if $\partial_u f(a)$ is defined and $\lambda \in \mathbb{R}^*$, then $\partial_{\lambda u} f(a)$ is defined and

$$\partial_{\lambda u} f(a) = \lambda \partial_u f(a).$$

If $E = \mathbb{R}^n$ and (e_i) is its standard basis, then the directional derivative $\partial_{e_i} f(a)$ is called the *i th partial derivative* of f at a, or the derivative of f with respect to x_i at a. In this case we write $\partial_i f(a)$ or $\frac{\partial f}{\partial x_i}(a)$. If $a = (a_1, \dots, a_n)$, then

$$\frac{\partial f}{\partial x_i}(a) = \lim_{t \to 0} \frac{f(a_1, \dots, a_i + t, \dots, a_n) - f(a_1, \dots, a_n)}{t}.$$

If for every point $x \in O$, the partial derivative $\frac{\partial f}{\partial x_i}(x)$ is defined, then we obtain the function ith partial derivative $\frac{\partial f}{\partial x_i}$ defined on O. If these functions are defined and continuous for all i, then we say that the function f is of *class C^1*.

R. Coleman, *Calculus on Normed Vector Spaces*, Universitext,
DOI 10.1007/978-1-4614-3894-6_2, © Springer Science+Business Media New York 2012

Example. If f is the function defined on $\mathbb{R}^2$ by $f(x, y) = xe^{xy}$, then the partial derivatives with respect to x and y are defined at all points $(x, y) \in \mathbb{R}^2$ and

$$\frac{\partial f}{\partial x}(x, y) = (1 + xy)e^{xy} \quad \text{and} \quad \frac{\partial f}{\partial y}(x, y) = x^2 e^{xy}.$$

As the functions $(x, y) \longmapsto (1 + xy)e^{xy}$ and $(x, y) \longmapsto x^2 e^{xy}$ are continuous, f is of class C^1.

Remark. If I is an open interval of $\mathbb{R}$, $a \in I$ and $f : I \longrightarrow \mathbb{R}$ has a derivative at a, then f is continuous at a. We have

$$\frac{f(a + t) - f(a)}{t} = \frac{df}{dt}(a) + \epsilon(t),$$

where $\lim_{t \to 0} \epsilon(t) = 0$. This implies that

$$f(a + t) - f(a) = t\frac{df}{dt}(a) + t\epsilon(t),$$

and the continuity of f at a follows. However, a function of two or more variables may have all its partial derivatives defined at a given point without being continuous there. Here is an example. Consider the function f defined on $\mathbb{R}^2$ by

$$f(x, y) = \begin{cases} \frac{x^6}{x^8 + (y - x^2)^2} & \text{if} \quad (x, y) \neq (0, 0) \\ 0 & \text{otherwise} \end{cases}.$$

We have

$$\lim_{t \to 0} \frac{t^6}{t^8 + t^4}/t = 0 \quad \text{and} \quad \lim_{t \to 0} \frac{0}{t^2}/t = 0$$

and so

$$\frac{\partial f}{\partial x}(0, 0) = \frac{\partial f}{\partial y}(0, 0) = 0.$$

However, $\lim_{x \to 0} f(x, x^2) = \infty$, which implies that f is not continuous at 0.

The next result needs no proof. It is simply an application of the definition of a partial derivative.

Proposition 2.1. *Let O be an open subset of $\mathbb{R}^n$, $a \in O$ and f and g real-valued functions defined on O having partial derivatives with respect to x_i at a. Then*

$$\frac{\partial(f + g)}{\partial x_i}(a) = \frac{\partial f}{\partial x_i}(a) + \frac{\partial g}{\partial x_i}(a) \quad and \quad \frac{\partial(fg)}{\partial x_i}(a) = \frac{\partial f}{\partial x_i}(a)g(a) + f(a)\frac{\partial g}{\partial x_i}(a).$$

In addition, if $\lambda \in \mathbb{R}$ then

$$\frac{\partial(\lambda f)}{\partial x_i}(a) = \lambda \frac{\partial(f)}{\partial x_i}(a).$$

Suppose now that O is an open subset of $\mathbb{R}^n$ and f a mapping defined on O with image in $\mathbb{R}^m$. f has m coordinate mappings $f_1, \ldots, f_m$. If $a \in O$ and the partial derivatives $\frac{\partial f_i}{x_j}(a)$, for $1 \leq i \leq m$ and $1 \leq j \leq n$, are all defined, then the $m \times n$ matrix

$$J_f(a) = \begin{pmatrix} \frac{\partial f_1}{\partial x_1}(a) & \ldots & \frac{\partial f_1}{\partial x_n}(a) \\ \vdots & \vdots & \vdots \\ \frac{\partial f_m}{\partial x_1}(a) & \ldots & \frac{\partial f_m}{\partial x_n}(a) \end{pmatrix}$$

is called the *Jacobian matrix* of f at a.

Example. If the mapping f of $\mathbb{R}^3$ into $\mathbb{R}^2$ is defined by $f(x, y, z) = (xy, ze^{xy})$, then all partial derivatives are defined at any point $(x, y, z) \in \mathbb{R}^3$ and

$$J_f(x, y, z) = \begin{pmatrix} y & x & 0 \\ yze^{xy} & xze^{xy} & e^{xy} \end{pmatrix}.$$

It is easy to generalize the definition of class C^1 to a mapping having its image in $\mathbb{R}^m$. We say that such a function is of class C^1 if its coordinate mappings are all of class C^1.

Remark. We do not need to restrict the directional derivative to functions defined on open sets of a normed vector space. Let A be a nonempty subset of a vector space E, f a real-valued function defined on A, and $a \in A$. If u is a nonzero element of E and there exists $\epsilon > 0$ such that $a + tu \in A$, when $|t| < \epsilon$, then the function $f_u : t \longrightarrow f(a + tu)$ is defined on the open interval $(-\epsilon, \epsilon)$. If the derivative $\frac{df_u}{dt}(0)$ is defined, then as above we note this derivative $\partial f_u(a)$ and call it the derivative of f at a in the direction u.

2.2 The Differential

Let E and F be normed vector spaces, O an open subset of E containing 0, and g a mapping from O into F such that $g(0) = 0$. If there exists a mapping ϵ, defined on a neighbourhood of $0 \in E$ and with image in F, such that $\lim_{h \to 0} \epsilon(h) = 0$ and

$$g(h) = \|h\|_E \epsilon(h),$$

then we write $g(h) = o(h)$ and say that g is a "small o of h". If $\| \cdot \|_E^\times \sim \| \cdot \|_E$ and $\| \cdot \|_F^\times \sim \| \cdot \|_F$, then $g(h) = o(h)$ for the norms $\| \cdot \|_E$ and $\| \cdot \|_F$ if and only if $g(h) = o(h)$ for the norms $\| \cdot \|_E^\times$ and $\| \cdot \|_F^\times$. In particular, if $E = \mathbb{R}^n$ and $F = \mathbb{R}^m$, then the condition $g(h) = o(h)$ is independent of the norms we choose for the two spaces.

Let O be an open subset of a normed vector space E and f a mapping from O into a normed vector space F. If $a \in O$ and there is a continuous linear mapping ϕ from E into F such that

$$f(a + h) = f(a) + \phi(h) + o(h)$$

when h is close to 0, then we say that f is *differentiable* at a.

Proposition 2.2. *If f is differentiable at a, then*

(a) *f is continuous at a;*
(b) *ϕ is unique.*

Proof. (a) As ϕ is continuous at 0, $\lim_{h \to 0} \phi(h) = 0$ and so $\lim_{h \to 0} f(a + h) = f(a)$.
(b) Suppose that

$$f(a + h) = f(a) + \phi_1(h) + o(h) = f(a) + \phi_2(h) + o(h)$$

and let $x \in E$. For $t > 0$ small we have

$$f(a + tx) - f(a) = t\phi_1(x) + t\|x\|_E \epsilon_1(tx) = t\phi_2(x) + t\|x\|_E \epsilon_2(tx),$$

where $\lim_{t \to 0} \epsilon_i(tx) = 0$. This implies that

$$\phi_2(x) - \phi_1(x) = \|x\|_E(\epsilon_1(tx) - \epsilon_2(tx))$$

Letting t go to 0, we obtain $\phi_1(x) - \phi_2(x) = 0$. $\square$

This unique continuous linear mapping ϕ is called the *differential* of f at a, written $f'(a)$, $df(a)$ or Df(a). If f is differentiable at every point $a \in O$, then we say that f is differentiable on O. If in addition f is a bijection onto an open subset U of F and the inverse mapping f^{-1} is also differentiable, then we say that f a *diffeomorphism*. Clearly a diffeomorphism is a homeomorphism.

Notation. We will use the notation f' for differentials. If considering the derivative of a real-valued function f defined on an open interval of $\mathbb{R}$, then we will use the notation $\frac{df}{dt}$ or $\dot{f}$. To simplify the notation, we will usually write $f'(a)h$ for $f'(a)(h)$.

Examples. If E and F are normed vector spaces and $f : E \longrightarrow F$ is constant, then $f'(a)$ is the zero mapping at any point $a \in E$. If $f : E \longrightarrow F$ is linear and continuous, then $f'(a) = f$ at any point $a \in E$.

Exercise 2.1. Let E be a normed vector space and f a real-valued function defined on E such that $|f(x)| \leq \|x\|_E^2$. Show that f is differentiable at 0.

Proposition 2.3. *If we replace the norms on the spaces E and F by equivalent norms, then the differentiability at $a \in O$ and the differential are unaffected. In particular, if E and F are finite-dimensional, then we may choose any pair of norms.*

Proof. If f is differentiable at a and

$$g(h) = f(a+h) - f(a) - f'(a)h,$$

then $g(h) = o(h)$. If we replace one or both the norms of E and F by an equivalent norm, then with respect to the new pair of norms, we have $g(h) = o(h)$. It follows that f is differentiable with respect to the second pair of norms and that the differential at a is the same. $\square$

Example. Let $A \in \mathcal{M}_n(\mathbb{R})$ be symmetric, $b \in \mathbb{R}^n$ and $f : \mathbb{R}^n \longrightarrow \mathbb{R}$ defined by

$$f(x) = \frac{1}{2}x^t A x - b^t x.$$

(Here, as elsewhere, when employing matrices, we identify elements of $\mathbb{R}^n$ with n-coordinate column vectors.) Let $a \in \mathbb{R}^n$. A simple calculation shows that

$$f(a+h) = f(a) + (a^t A - b^t)h + \frac{1}{2}h^t A h.$$

The function $\phi : h \longmapsto (a^t A - b^t)h$ is linear. As $\mathbb{R}^n$ is finite-dimensional, ϕ is also continuous. We also have

$$|h^t A h| \leq \|Ah\|_2 \|h\|_2 \leq |A|_2 \|h\|_2^2,$$

where $| \cdot |_2$ is the matrix norm subordinate to the norm $\| \cdot \|_2$. Hence $f'(a) = \phi$.

Exercise 2.2. Show that the mapping

$$f : \mathcal{M}_n(\mathbb{R}) \longrightarrow \mathcal{M}_n(\mathbb{R}), \ X \longmapsto X^t X$$

is differentiable at any point $A \in \mathcal{M}_n(\mathbb{R})$ and determine $f'(A)$.

Proposition 2.4. *Let f be a mapping defined on an open subset O of a normed vector space E with image in the cartesian product $F = F_1 \times \cdots \times F_p$. Then f is differentiable at $a \in O$ if and only if the coordinate mappings f_i, for $i = 1, \ldots, p$, are differentiable at a.*

Proof. Suppose first that the coordinate mappings are differentiable at a:

$$f_i(a+h) = f_i(a) + f_i'(a)h + \|h\|_E \epsilon_i(h),$$

where $\lim_{h \to 0} \epsilon_i(h) = 0$. The mapping

$$\phi : E \longrightarrow F, h \longmapsto (f_1'(a)h, \ldots, f_p'(a)h)$$

is a continuous linear mapping. If we set $\epsilon(h) = (\epsilon_1(h), \ldots, \epsilon_p(h))$, then $\lim_{h \to 0} \epsilon(h) = 0$ and

$$f(a + h) = f(a) + \phi(h) + \|h\|_E \epsilon(h).$$

Therefore f is differentiable at a.

Suppose now that f is differentiable at a:

$$f(a + h) = f(a) + f'(a)h + \|h\|_E \epsilon(h),$$

where $\lim_{h \to 0} \epsilon(h) = 0$. Then

$$f_i(a + h) = f_i(a) + L_i(h) + \|h\|_E \epsilon_i(h),$$

where $f'(a)h = (L_1(h), \ldots, L_p(h))$. For each i the mapping L_i is linear and continuous and $\lim_{h \to 0} \epsilon_i(h) = 0$; hence f_i is differentiable at a. $\qquad \square$

Remark. The differential $f'(a)$ is a continuous linear mapping from E into F. In the above proof we have shown that the coordinate mappings of $f'(a)$ at a are the differentials at a of the coordinate mappings of f, i.e.,

$$f'(a) = (f_1'(a), \ldots, f_p'(a)).$$

Proposition 2.5. *If O is an open subset of a normed vector space E and $f : E \longrightarrow \mathbb{R}$ is differentiable at $a \in O$, then the directional derivative $\partial f_u(a)$ is defined for any nonzero vector $u \in E$ and $\partial f_u(a) = f'(a)u$. In particular, if $E = \mathbb{R}^n$, then the partial derivatives $\frac{\partial f}{\partial x_1}(a), \ldots, \frac{\partial f}{\partial x_n}(a)$ are defined.*

Proof. We have

$$f(a + tu) = f(a) + tf'(a)u + o(tu)$$

and so

$$\lim_{t \to 0} \frac{f(a + tu) - f(a)}{t} = f'(a)u.$$

This ends the proof. $\qquad \square$

Suppose that $\dim E = n < \infty$ and that (e_i) is a basis of E. If $x = \sum_{i=1}^{n} x_i e_i$, then

$$f'(a)x = \sum_{i=1}^{n} x_i f'(a)e_i = \sum_{i=1}^{n} \partial_{e_i} f(a) e_i^*(x),$$

where (e_i^*) is the dual basis of (e_i). We thus obtain the expression

$$f'(a) = \sum_{i=1}^{n} \partial_{e_i} f(a) e_i^*.$$

If $E = \mathbb{R}^n$ and (e_i) its standard basis, then we usually write dx_i for e_i^*. This gives us the expression

$$f'(a) = \sum_{i=1}^{n} \frac{\partial f}{\partial x_i}(a) dx_i.$$

What we have just seen has practical importance. If we wish to determine whether a real-valued function f defined on an open subset of $\mathbb{R}^n$ is differentiable at a given point a, then first we determine whether all its partial derivatives at a exist. If this is not the case, then f is not differentiable at a. If all the partial derivatives exist, then we know that the only possibility for $f'(a)$ is the linear function $\phi = \sum_{i=1}^{n} \frac{\partial f}{\partial x_i}(a) dx_i$. To conclude, we consider the expression

$$\frac{f(a+h) - f(a) - \phi(h)}{\|h\|} = \epsilon(h).$$

If $\lim_{h\to 0} \epsilon(h) = 0$, then f is differentiable at a, otherwise it is not.

Example. Let $f : \mathbb{R}^2 \longrightarrow \mathbb{R}$ be defined by

$$f(x, y) = \begin{cases} \frac{x^3 - y^3}{x^2 + y^2} & \text{if } (x, y) \neq (0, 0) \\ 0 & \text{otherwise} \end{cases}.$$

A simple calculation shows that

$$\frac{\partial f}{\partial x}(0, 0) = 1 \quad \text{and} \quad \frac{\partial f}{\partial y}(0, 0) = -1.$$

Therefore, if $f'(0, 0)$ exists, then $f'(0, 0) = dx - dy$. However,

$$\left| \frac{h^3 - k^3}{h^2 + k^2} - (h - k) \right| / \|(h, k)\|_2 = \frac{|hk(h - k)|}{(h^2 + k^2)^{\frac{3}{2}}}.$$

Setting $k = -h$ in the expression on the right-hand side of the equation, we obtain $\frac{1}{\sqrt{2}}$. Therefore we do not have the necessary convergence and so f is not differentiable at $(0, 0)$.

The previous example shows that a function may have partial derivatives at a point without being differentiable at that point. This however is not the case for functions of a single variable.

Proposition 2.6. *Let I be an open interval of* $\mathbb{R}$. *Then* $f : I \longrightarrow \mathbb{R}$ *is differentiable at* $a \in I$ *if and only if* f *has a derivative at* a.

Proof. From Proposition 2.5, if f is differentiable at a, then f has a derivative at a. Now suppose that $\frac{\mathrm{d}f}{\mathrm{d}x}(a)$ exists. Then

$$\frac{f(a+h) - f(a)}{h} - \frac{\mathrm{d}f}{\mathrm{d}x}(a) = \epsilon(h),$$

where $\lim_{h \to 0} \epsilon(h) = 0$. Multiplying by h we obtain

$$f(a+h) = f(a) + \frac{\mathrm{d}f}{\mathrm{d}x}(a)h + h\epsilon(h).$$

It follows that f is differentiable at a and that $f'(a)h = \frac{\mathrm{d}f}{\mathrm{d}x}(a)h$. □

If O is an open subset of $\mathbb{R}^n$ and $f : O \longrightarrow \mathbb{R}^m$ is differentiable at $a \in O$, then $f'(a)$ is a linear mapping from $\mathbb{R}^n$ into $\mathbb{R}^m$. Let us write (e_j^n) (resp.(e_i^m)) for the standard basis of $\mathbb{R}^n$ (resp. $\mathbb{R}^m$). We have $f'(a) = (f_1'(a), \ldots, f_m'(a))$. However, $f_i'(a)e_j^n = \frac{\partial f_i}{\partial x_j}(a)$ and so

$$f'(a)e_j^n = \left(\frac{\partial f_1}{\partial x_j}(a), \ldots, \frac{\partial f_m}{\partial x_j}(a) \right) = \sum_{i=1}^{m} \frac{\partial f_i}{\partial x_j}(a)e_i^m.$$

Therefore the jth column of the matrix of $f'(a)$ with respect to the bases (e_j^n) and (e_i^m) has for elements $\frac{\partial f_1}{\partial x_j}(a), \ldots, \frac{\partial f_m}{\partial x_j}(a)$. It follows that this matrix is the Jacobian matrix $J_f(a)$.

Notation. If E and F are finite-dimensional vector spaces, $\mathcal{B}_E$ and $\mathcal{B}_F$ bases of these spaces and $L : E \longrightarrow F$ a linear mapping, then we will write $\mathrm{mat}_{\mathcal{B}_E \mathcal{B}_F} L$ for the matrix of L with respect to the bases $\mathcal{B}_E$ and $\mathcal{B}_F$. So we could write

$$\mathrm{mat}_{(e_i^n)(e_i^m)} f'(a) = J_f(a).$$

The proof of the next result is elementary.

Proposition 2.7. *Let E and F be normed vector spaces, O an open subset of E and a an element of O. If* f *and* g *are differentiable at* a, *then* $f + g$ *is differentiable at* a, *as is* λf, *for any* $\lambda \in \mathbb{R}$, *and*

$$(f + g)'(a) = f'(a) + g'(a) \qquad \text{and} \qquad (\lambda f)'(a) = \lambda(f'(a)).$$

If F is a commutative normed algebra, then fg *is differentiable at* a *and*

$$(fg)'(a) = f(a)g'(a) + g(a)f'(a).$$

Suppose that E and F are normed vector spaces, O an open subset of E and $f : O \longrightarrow F$ differentiable at $a \in O$. If $\tilde{F}$ is a vector subspace of F and the image of f lies in $\tilde{F}$, then f is differentiable at a as a mapping from O into $\tilde{F}$ if the image of $f'(a)$ lies in $\tilde{F}$. The following result gives us a sufficient condition for this to be so.

Proposition 2.8. *If $\tilde{F}$ is closed, then f is differentiable at a as a mapping from O into $\tilde{F}$.*

Proof. For any $h \in E$

$$\lim_{t \to 0} \frac{f(a + th) - f(a)}{t} = f'(a)h.$$

Let (t_n) be a sequence in $\mathbb{R}^*$ with limit 0 and such that $a + t_n h \in O$. If we set $u_n = \frac{f(a + t_n h) - f(a)}{t_n}$, then the sequence (u_n) is a convergent sequence contained in $\tilde{F}$. As $\tilde{F}$ is closed, its limit $f'(a)h$ is an element this subspace. This ends the proof. $\square$

2.3 Differentials of Compositions

In this section we consider the differentiability of mappings which are compositions. Let E, F and G be normed vector spaces, O an open subset of E, U an open subset of F and $f : O \longrightarrow F$, $g : U \longrightarrow G$ be such that $f(O) \subset U$. Then the mapping $g \circ f$ is defined on O.

Theorem 2.1. *If f is differentiable at a and g is differentiable at $f(a)$, then $g \circ f$ is differentiable at a and*

$$(g \circ f)'(a) = g'(f(a)) \circ f'(a).$$

Proof. To simplify the notation let us write $b = f(a)$. We have

$$f(a + h) = f(a) + f'(a)h + \|h\|_E \epsilon_1(h)$$

and

$$g(b + k) = g(b) + g'(b)k + \|k\|_F \epsilon_2(k),$$

with $\lim_{h \to 0} \epsilon_1(h) = 0$ and $\lim_{k \to 0} \epsilon_2(k) = 0$. For h sufficiently small we may write

$$\begin{aligned}
g(f(a + h)) &= g(b + f'(a)h + \|h\|_E \epsilon_1(h)) \\
&= g(b) + g'(b)(f'(a)h + \|h\|_E \epsilon_1(h)) \\
&\quad + \|f'(a)h + \|h\|_E \epsilon_1(h)\|_F \epsilon_2(f'(a)h + \|h\|_E \epsilon_1(h)) \\
&= g(b) + g'(b) \circ f'(a)h + \|h\|_E \tilde{\epsilon}(h),
\end{aligned}$$

where

$$\tilde{\epsilon}(h) = g'(b)\epsilon_1(h) + \|f'(a)\frac{h}{\|h\|_E} + \epsilon_1(h)\|_F \epsilon_2\big(f'(a)h + \|h\|_E \epsilon_1(h)\big).$$

To finish, we only need to show that $\lim_{h\to 0} \tilde{\epsilon}(h) = 0$. However,

- $\lim_{h\to 0}\epsilon_1(h) = 0 \implies \lim_{h\to 0} g'(b)\epsilon_1(h)$, because $g'(b)$ is continuous;
- $\|f'(a)\frac{h}{\|h\|_E} + \epsilon_1(h)\|_F$ is bounded for small values of h, because $f'(a)$ is continuous and so bounded on the unit sphere;
- $\lim_{h\to 0}\epsilon_2(f'(a)h + \|h\|_E \epsilon_1(h)) = 0$, because $f'(a)$ is continuous.

Therefore $\lim_{h\to 0} \tilde{\epsilon}(h) = 0$. $\square$

Corollary 2.1. *If in the above theorem the normed vector spaces are euclidian spaces, then*

$$J_{g\circ f}(a) = J_g(f(a))J_f(a).$$

Proof. If $E = \mathbb{R}^n$, $F = \mathbb{R}^m$, $G = \mathbb{R}^s$ and $\mathcal{B}^n$, $\mathcal{B}^m$, $\mathcal{B}^s$ their standard bases, then we have

$$
\begin{aligned}
J_{g\circ f}(a) &= \text{mat }_{\mathcal{B}^n \mathcal{B}^s}(g\circ f)'(a) \\
&= \text{mat }_{\mathcal{B}^n \mathcal{B}^s} g'(f(a)) \circ f'(a) \\
&= \text{mat }_{\mathcal{B}^m \mathcal{B}^s} g'(f(a))\text{mat }_{\mathcal{B}^n \mathcal{B}^m} f'(a) \\
&= J_g(f(a))J_f(a),
\end{aligned}
$$

which is the result we were looking for. $\square$

Remark. The expression

$$(g\circ f)'(a) = g'(f(a)) \circ f'(a).$$

is often referred to as the *chain rule*.

Example. Let $f : \mathbb{R}^3 \longrightarrow \mathbb{R}^2$ and $g : \mathbb{R}^2 \longrightarrow \mathbb{R}$ be defined by

$$f(x, y, z) = (xy, e^{xz}) \qquad \text{and} \qquad g(u, v) = u^2 v.$$

Then

$$J_f(x, y, z) = \begin{pmatrix} y & x & 0 \\ ze^{xz} & 0 & xe^{xz} \end{pmatrix} \qquad \text{and} \qquad J_g(u, v) = \begin{pmatrix} 2uv & u^2 \end{pmatrix}.$$

Setting $u = xy$ and $v = e^{xz}$ in the second matrix, we obtain

$$J_g(f(x, y, z)) = \begin{pmatrix} 2xye^{xz} & x^2 y^2 \end{pmatrix}.$$

Multiplying the matrices $J_g(f(x, y, z))$ and $J_f(x, y, z)$, we obtain

$$J_{g \circ f}(x, y, z) = \left((2xy^2 + x^2y^2z)e^{xz} \quad 2x^2ye^{xz} \quad x^3y^2e^{xz} \right).$$

Exercise 2.3. Let I be an open interval of $\mathbb{R}$, O an open subset of $\mathbb{R}^m$, f a mapping of I into $\mathbb{R}^m$ and g a real-valued function defined on O. We suppose that $f(I) \subset O$. If $a \in I$ and $f'(a)$ and $g'(f(a))$ exist, show that

$$\frac{\mathrm{d}}{\mathrm{d}t} g \circ f(t) = \sum_{i=1}^{m} \frac{\partial g}{\partial x_i}(f(t)) \frac{\mathrm{d}f_i}{\mathrm{d}t}(t).$$

For the functions

$$f : \mathbb{R} \longrightarrow \mathbb{R}^2, t \longmapsto (t^2, t^3) \qquad \text{and} \qquad g : \mathbb{R}^2 \longrightarrow \mathbb{R}, (x, y) \longmapsto xy$$

calculate $\frac{\mathrm{d}}{\mathrm{d}t} g \circ f(t)$ using the formula. Find an expression for the function $g \circ f$ and then confirm the result.

Exercise 2.4. Let E and F be normed vector spaces, O an open subset of E, U an open subset of F and $f : O \longrightarrow U$ a diffeomorphism. Show that, for any point $x \in O$, $f'(x)$ is a normed vector space isomorphism from E into F and that $(f^{-1})'(f(x)) = f'(x)^{-1}$.

2.4   Mappings of Class C^1

If $O \subset \mathbb{R}^n$ is open and f a real-valued function whose partial derivatives are defined and continuous on O, then we say that f is of class C^1. More generally, if $f : O \longrightarrow \mathbb{R}^m$ is such that the mn functions $\frac{\partial f_i}{\partial x_j}$ are defined and continuous on O, then we say that f is of *class* C^1. In this section we will obtain a result giving necessary and sufficient conditions for a mapping to be of class C^1. This will enable us to generalize this notion to mappings between any pair of normed vector spaces.

Consider a real-valued function f defined on an open subset O of $\mathbb{R}^n$. We have seen that if f has partial derivatives at a point $a \in O$, then this does not imply that f is differentiable at a. However, if we add a condition, then we do obtain differentiability.

Theorem 2.2. *If the functions $\frac{\partial f}{\partial x_1}, \ldots, \frac{\partial f}{\partial x_n}$ are defined on a neighbourhood V of a and continuous at a, then f is differentiable at a.*

Proof. As V is a neighbourhood of a, there is an open cube

$$C(a, \epsilon) = \{x \in \mathbb{R}^n : |x_i - a_i| < \epsilon\}$$

contained in V. If h lies in the open cube $C(0, \epsilon)$, then $a + h \in C(a, \epsilon)$ and

$$
\begin{aligned}
f(a + h) - f(a) &= f(a_1 + h_1, \ldots, a_n + h_n) - f(a_1, \ldots, a_n) \\
&= f(a_1 + h_1, \ldots, a_n + h_n) - f(a_1 + h_1, \ldots, a_{n-1} + h_{n-1}, a_n) \\
&\quad + f(a_1 + h_1, \ldots, a_{n-1} + h_{n-1}, a_n) \\
&\quad - f(a_1 + h_1, \ldots, a_{n-2} + h_{n-2}, a_{n-1}, a_n) \\
&\quad \vdots \\
&\quad + f(a_1 + h_1, a_2, \ldots, a_n) - f(a_1, \ldots, a_n).
\end{aligned}
$$

Suppose that $h_n \neq 0$ and consider the function

$$
g_n : [0, h_n] \longrightarrow \mathbb{R}, t \longmapsto f(a_1 + h_1, \ldots, a_{n-1} + h_{n-1}, a_n + t).
$$

From the mean value theorem there exists $\theta \in (0, 1)$ such that

$$
g_n(h_n) - g_n(0) = \frac{dg_n}{dt}(\theta h_n) h_n.
$$

It follows that there exists $y^n \in C(a, \epsilon)$ such that

$$
f(a_1 + h_1, \ldots, a_n + h_n) - f(a_1 + h_1, \ldots, a_{n-1} + h_{n-1}, a_n) = \frac{\partial f}{\partial x_n}(y^n) h_n.
$$

(If $h_n = 0$, then we may take $y^n = a$.) We proceed in the same way for each line on the right-hand side of the above expression to obtain $y^1, \ldots, y^n \in C(a, \epsilon)$ such that

$$
f(a + h) - f(a) = \sum_{i=1}^{n} h_i \frac{\partial f}{\partial x_i}(y^i).
$$

Therefore

$$
f(a + h) - f(a) - \sum_{i=1}^{n} h_i \frac{\partial f}{\partial x_i}(a) = \sum_{i=1}^{n} h_i \left(\frac{\partial f}{\partial x_i}(y^i) - \frac{\partial f}{\partial x_i}(a) \right).
$$

Now,

$$
\left| \sum_{i=1}^{n} h_i \left(\frac{\partial f}{\partial x_i}(y^i) - \frac{\partial f}{\partial x_i}(a) \right) \right| \leq \|h\|_\infty \sum_{i=1}^{n} \left| \frac{\partial f}{\partial x_i}(y^i) - \frac{\partial f}{\partial x_i}(a) \right|
$$

and $\lim_{h \to 0} y^i = a$ for each i. As the functions $\frac{\partial f}{\partial x_i}$ are continuous at a, we have

$$\lim_{h \to 0} \sum_{i=1}^{n} \left| \frac{\partial f}{\partial x_i}(y^i) - \frac{\partial f}{\partial x_i}(a) \right| = 0.$$

It follows that

$$f(a+h) - f(a) - \sum_{i=1}^{n} h_i \frac{\partial f}{\partial x_i}(a) = o(h)$$

and so f is differentiable at a. $\qquad\square$

Corollary 2.2. *Let O be an open subset of $\mathbb{R}^n$ and $f : O \longrightarrow \mathbb{R}^m$ such that the functions $\frac{\partial f_i}{\partial x_j}$, for $1 \le i \le m$ and $1 \le j \le n$, are defined on a neighbourhood V of $a \in O$ and continuous at a. Then f is differentiable at a.*

Proof. From the theorem above each coordinate function is differentiable at a and so f is differentiable at a. $\qquad\square$

Examples. 1. If $f : \mathbb{R}^2 \longrightarrow \mathbb{R}^2$ is defined by

$$f(r, \theta) = (r \cos \theta, r \sin \theta),$$

then we have

$$\frac{\partial f_1}{\partial r} = \cos \theta, \quad \frac{\partial f_1}{\partial \theta} = -r \sin \theta, \quad \frac{\partial f_2}{\partial r} = \sin \theta \quad \text{and} \quad \frac{\partial f_2}{\partial \theta} = r \cos \theta.$$

The four partial derivatives are clearly continuous; therefore f is differentiable at any point (r, θ) and

$$f'(r, \theta) = (\cos \theta dr - r \sin \theta d\theta, \sin \theta dr + r \cos \theta d\theta).$$

2. If $f : \mathbb{R}^n \longrightarrow \mathbb{R}$ is defined by

$$f(x_1, \ldots, x_n) = x_1 x_2 \cdots x_n,$$

then

$$\frac{\partial f}{\partial x_i}(x) = x_1 \cdots x_{i-1} \hat{x}_i x_{i+1} \cdots x_n,$$

where the "hat" indicates that the variable is absent. The n partial derivatives are clearly continuous; hence f is differentiable at any point x and

$$f'(x) = \sum_{i=1}^{n} x_1 \cdots x_{i-1} \hat{x}_i x_{i+1} \cdots x_n dx_i.$$

It is relatively easy to extend Theorem 2.2 to general finite-dimensional normed vector spaces. Let E be an n-dimensional normed vector space, (v_i) a basis of E, (e_i) the standard basis of $\mathbb{R}^n$ and L the linear mapping which sends v_i to e_i. Consider a real-valued function f defined on an open subset O of E and suppose that the directional derivatives $\partial_{v_i} f(x)$ are defined and continuous on a neighbourhood V of a point $a \in O$. If we set $\tilde{f} = f \circ L^{-1}$, then $\tilde{f}$ is defined on the open subset $L(O)$ of $\mathbb{R}^n$ and $L(V)$ is a neighbourhood of $L(a)$. A simple calculation shows that $\frac{\partial \tilde{f}}{\partial x_i}(L(x)) = \partial_{v_i} f(x)$ and it follows that $\tilde{f}$ is differentiable at $L(a)$. As L is differentiable at a and $f = \tilde{f} \circ L$, f is differentiable at a and

$$f'(a) = \tilde{f}'(L(a)) \circ L'(a) = \tilde{f}'(L(a)) \circ L.$$

Example. Suppose that $n \geq 1$ and let us consider the mapping $\det : \mathcal{M}_n(\mathbb{R}) \longrightarrow \mathbb{R}$, where $\det(X)$ is the determinant of X. We will write $E(i, j)$, with $1 \leq i, j \leq n$, for the elements of the standard basis of $\mathcal{M}_n(\mathbb{R})$, i.e.,

$$E(i, j)_{k,l} = \begin{cases} 1 & \text{if } i = k, j = l \\ 0 & \text{otherwise} \end{cases}.$$

For $X \in \mathcal{M}_n(\mathbb{R})$ we have

$$\det(X + t E(i, j)) - \det(X) = t \gamma_{ij}(X),$$

where $\gamma_{ij}(X)$ is the (i, j)-cofactor of X. It follows that

$$\partial_{E(i,j)} \det X = \gamma_{ij}(X).$$

As $\gamma_{ij}(X)$ is a polynomial in the entries X_{ij} of X, $\partial_{E(i,j)} X$ is a continuous function of X and so the mapping $\det$ is differentiable at any point $X \in \mathcal{M}_n(\mathbb{R})$.

Proposition 2.9. *Let O be an open subset of $\mathbb{R}^n$, f and g real-valued functions of class C^1 defined on O and $\lambda \in \mathbb{R}$. Then $f + g$, λf and fg are of class C^1. If g does not vanish on O, then $\frac{f}{g}$ is of class C^1.*

Proof. We have the relations

$$\frac{\partial(f + g)}{\partial x_i}(x) = \frac{\partial f}{\partial x_i}(x) + \frac{\partial g}{\partial x_i}(x) \qquad \text{and} \qquad \frac{\partial(\lambda g)}{\partial x_i}(x) = \lambda \frac{\partial f}{\partial x_i}(x).$$

Also,
$$\frac{\partial(fg)}{\partial x_i}(x) = \frac{\partial f}{\partial x_i}(x)g(x) + f(x)\frac{\partial g}{\partial x_i}(x).$$

As f and g are of class C^1, the functions f, g, $\frac{\partial f}{\partial x_i}$ and $\frac{\partial g}{\partial x_i}$ are continuous; hence the right-hand sides of the above expressions are continuous and so the functions $f + g$, λf and fg are of class C^1. If g does not vanish and we set $h = \frac{f}{g}$, then

$$\frac{\partial h}{\partial x_i}(x) = \frac{g(x)\frac{\partial f}{\partial x_i}(x) - f(x)\frac{\partial g}{\partial x_i}(x)}{(g(x))^2},$$

and so $\frac{\partial h}{\partial x_i}$ is continuous. Thus h is of class C^1.   □

Corollary 2.3. *Let O be an open subset of $\mathbb{R}^n$, f and g mappings of class C^1 defined on O with image in $\mathbb{R}^m$ and $\lambda \in \mathbb{R}$. Then $f + g$ and λf are of class C^1.*

Proof. As f and g are of class C^1, so are their coordinate mappings. Now

$$(f + g)_i = f_i + g_i \qquad \text{and} \qquad (\lambda f)_i = \lambda f_i$$

and so the coordinate mappings of $f + g$ and λf are of class C^1. It follows that $f + g$ and λf are C^1-mappings.   □

Remark. From the above proposition and corollary, we see that the C^1-mappings defined on O form a vector space if the image space is $\mathbb{R}^m$ and, in the case where $m = 1$, this is an algebra.

Proposition 2.10. *Let O be an open subset of $\mathbb{R}^n$, U an open subset of $\mathbb{R}^m$, f a mapping from O into $\mathbb{R}^m$ and g a mapping from U into $\mathbb{R}^s$. If $f(O) \subset U$ and f and g are of class C^1, then $g \circ f$ is of class C^1.*

Proof. For $x \in O$ we have

$$J_{g \circ f}(x) = J_g(f(x)) \circ J_f(x),$$

therefore

$$\frac{\partial (g \circ f)_i}{\partial x_j} = \sum_{k=1}^{m} \frac{\partial g_i}{\partial y_k}(f(x)) \frac{\partial f_k}{\partial x_j}(x).$$

As f and the partial derivatives $\frac{\partial g_i}{\partial y_k}$ and $\frac{\partial f_k}{\partial x_j}$ are continuous, the partial derivative $\frac{\partial (g \circ f)_i}{\partial x_j}$ is continuous. It follows that $g \circ f$ is of class C^1.   □

We now prove the characterization of C^1-mappings referred to at the beginning of the section.

Theorem 2.3. *Let O be an open subset of $\mathbb{R}^n$ and f a mapping from O into $\mathbb{R}^m$. Then f is of class C^1 if and only if f is differentiable on O and the mapping f' from O into $\mathcal{L}(\mathbb{R}^n, \mathbb{R}^m)$ is continuous.*

Proof. Let us set $E = \mathbb{R}^n$, $F = \mathbb{R}^m$ and fix norms on these spaces. Suppose that f is of class C^1. From Theorem 2.2, we know that f is differentiable on O. Also, for $x \in O$ and $u \in \mathbb{R}^n$ we have

$$f'(x)u = J_f(x)u.$$

Now

$$|f'(x)|_{\mathcal{L}(E,F)} = |J_f(x)|,$$

where $|\cdot|$ is the norm defined on $\mathcal{M}_{mn}(\mathbb{R})$ by

$$|A| = \sup_{\|x\|_E \le 1} \|Ax\|_F.$$

If we let

$$|A|_M = \max |a_{ij}|,$$

where $A = (a_{ij})$, then $|\cdot|_M$ also defines a norm on $\mathcal{M}_{mn}(\mathbb{R})$. As f is of class C^1, we have

$$\lim_{h \to 0} |\mathrm{J}_f(x + h) - \mathrm{J}_f(x)|_M = 0.$$

However, the norm $|\cdot|_M$ is equivalent to the norm $|\cdot|$ and so

$$\lim_{h \to 0} |\mathrm{J}_f(x + h) - \mathrm{J}_f(x)| = 0.$$

It follows that f' is continuous at x.

Now suppose that f' is defined and continuous on O. From Proposition 2.5 we know that the partial derivatives of f are defined on O. Also, for $x \in O$ and $u \in \mathbb{R}^n$ we have

$$f'(x)u = \mathrm{J}_f(x)u.$$

As f' is continuous, we have

$$\lim_{h \to 0} |\mathrm{J}_f(x + h) - \mathrm{J}_f(x)| = 0 \implies \lim_{h \to 0} |\mathrm{J}_f(x + h) - \mathrm{J}_f(x)|_M = 0,$$

which implies that the partial derivatives are continuous at x.   □

The preceding theorem suggests the following generalization of the notion of class C^1. If E and F are normed vector spaces and O is open in E, then $f : O \longrightarrow F$ is of *class C^1* if f is differentiable at every point $x \in O$ and the mapping $f' : O \longrightarrow \mathcal{L}(E, F)$ is continuous.

In closing this section we mention that the exponential mapping is of class C^1. This can be shown directly. However, the calculations are rather long. Later on we will prove this result in a simple way, avoiding arduous calculations.

2.5 Extrema

Let us recall the notions of minimum and maximum. Suppose that f is a real-valued function defined on a set X. We say that $a \in X$ is a *(global) minimum* if, for all $x \in X$,

$$f(a) \le f(x);$$

$a \in X$ is *(global) maximum* if, for all $x \in X$,

$$f(a) \geq f(x).$$

If the inequality is strict when $x \neq a$, then we speak of a *strict minimum* or *strict maximum*. A point which is either a minimum or a maximum is called an *extremum*. A function may have no minimum, a single minimum or several minima. The same is true for maxima.

Examples. 1. $f : \mathbb{R} \longrightarrow \mathbb{R}, x \longmapsto x$ has neither a minimum nor a maximum.
2. $f : \mathbb{R} \longrightarrow \mathbb{R}, x \longmapsto x^2$ has a minimum but no maximum.
3. $f : \mathbb{R} \longrightarrow \mathbb{R}, x \longmapsto -x^2$ has a maximum but no minimum.
4. $f : \mathbb{R} \longrightarrow \mathbb{R}, x \longmapsto \cos x$ has an infinite number of minima and maxima.

Let us now turn to normed vector spaces. We have already seen that if E is a normed vector space, $X \subset E$ compact and $f : X \longrightarrow \mathbb{R}$ continuous, then f has a minimum and a maximum (Theorem 1.3). This result is one of existence: it does not tell us how to find the extrema. We now introduce a related notion which can often help us in this direction. Let X be a subset of a normed vector space E and f a real-valued function defined on X. We say that $a \in X$ is a *local minimum* if a has a neighbourhood N such that

$$f(a) \leq f(x)$$

for all $x \in N \cap X$. We define a *local maximum* in an analogous way (reversing the direction of the inequality). As above, if the inequality is strict when $x \neq a$, then we speak of a strict local minimum or maximum. A point which is either a local minimum or maximum is called a local extremum . Clearly a global minimum (resp. maximum) is a local minimum (resp. maximum); however, the converse is not true. As a first step in looking for an extremum, it can be useful to look for local extrema. We will present a fundamental result which helps us to do so.

Let a and b be elements of a vector space E. We call the set

$$[a, b] = \{x \in E : x = \lambda a + (1 - \lambda)b , \lambda \in [0, 1]\}$$

the *segment* joining a to b. We write (a, b) for $[a, b] - \{a, b\}$. If $X \subset E$ is such that the segment $[a, b]$ always lies in E when $a, b \in E$, then we say that X is *convex*.

Exercise 2.5. Show that segments and affine subspaces are convex and that in a normed vector space closed and open balls are convex.

Theorem 2.4. *Let E be a normed vector space, $O \subset E$ open and $X \subset O$ convex and suppose that f is a real-valued function defined on O. If f restricted to X has a local minimum at x and f is differentiable at x, then*

$$f'(x)(y - x) \geq 0$$

for all $y \in X$.

Proof. Let $\lambda \in (0,1)$. As X is convex, $x + \lambda(y-x) \in X$ and for λ sufficiently small we have

$$f(x+\lambda(y-x)) - f(x) = f'(x)(\lambda(y-x)) + o(\lambda(y-x))$$
$$= \lambda f'(x)(y-x) + |\lambda| \|y-x\| \epsilon(\lambda(y-x)),$$

where $\lim_{h\to 0} \epsilon(h) = 0$. Dividing by λ, we obtain

$$f'(x)(y-x) + \frac{|\lambda|}{\lambda} \|y-x\| \epsilon(\lambda(y-x)) \geq 0,$$

because x is a local minimum. Letting λ go to 0, we obtain the result. $\square$

Remark. The above inequality is called *Euler's inequality*. If x is a maximum, then x is a minimum of the function $-f$ restricted to X and so we obtain the inequality in the opposite direction.

Corollary 2.4. *Let O be an open subset of a normed vector space E and X an affine subspace of E: $X = a + V$, where V is a vector subspace of E and $a \in X$. We suppose that $O \cap X \neq \emptyset$. If f is a real-valued function defined on O such that f restricted to X has a local extremum at x and f is differentiable at x, then*

$$f'(x)v = 0$$

for all $v \in V$.

Proof. We will prove the result for a local minimum; the other case can be proved by considering $-f$. Suppose that $f'(x)v \neq 0$ for some $v \in V$. Replacing v by $-v$ if necessary, we may suppose that $f'(x)v > 0$. Let B be an open ball centred on x and lying in O. If we set $X_1 = B \cap X$, then X_1 is a convex subset of O. There exists $\lambda > 0$ such that $y = x - \lambda v \in X_1$ and

$$f'(x)(y-x) = -\lambda f'(x)v \geq 0 \Longrightarrow f'(x)v \leq 0,$$

a contradiction. Hence $f'(x)v = 0$ for all $v \in V$. $\square$

Remarks. 1. A special case of the above result is when the function f is defined on E: if $f_{|X}$ has a local extremum at x, then

$$f'(x)v = 0$$

for all $v \in V$.
2. If $E = \mathbb{R}^n$, then we have
$$\nabla f(x)v = 0$$

for all $v \in V$, i.e., $\nabla f(x) \in V^\perp$.

Corollary 2.5. *Suppose that O is an open subset of a normed vector space E. If $f : O \longrightarrow \mathbb{R}$ has a local extremum at $x \in O$ and f is differentiable at x, then $f'(x) = 0$.*

Proof. This result follows immediately from Corollary 2.4: it is sufficient to set $X = E$.   □

Suppose that O is an open subset of a normed vector space E and that the mapping $f : O \longrightarrow \mathbb{R}$ is differentiable at a point $x \in O$. If $f'(x) = 0$, then we say that x is a *critical point* of f. From what we have just seen, if x is a local extremum, then x is a critical point; however, the converse is false. For example, 0 is a critical point of the mapping $f : \mathbb{R} \longrightarrow \mathbb{R}, x \longmapsto x^3$, but 0 is not a local extremum.

Example. Consider the function

$$f : \mathbb{R}^2 \longrightarrow \mathbb{R}, (x, y) \longmapsto x^3 y^2 (1 - x - y)$$

defined on the set

$$X = \left\{ (x, y \in \mathbb{R}^2 : x \geq 0, y \geq 0, x + y \leq 1 \right\}.$$

As X is closed and bounded, X is compact. Therefore f has a minimum and a maximum on X. Clearly $f(x, y) \geq 0$ and $f(x, y) = 0$ if and only if $x = 0, y = 0$ or $x + y = 1$, i.e., if and only if (x, y) lies on the boundary of X. Hence the minima of f are those points lying on the boundary of X. As there are points (x, y) such that $f(x, y) > 0$, any maximum (x, y) of f must lie in the interior O of X, an open set. Such a point is a maximum of the function restricted to O and so a critical point of this function. We have

$$\frac{\partial f}{\partial x}(x, y) = x^2 y^2 (3 - 4x - 3y) \qquad \text{and} \qquad \frac{\partial f}{\partial y}(x, y) = x^3 y(2 - 2x - 3y).$$

Setting the partial derivatives equal to 0, we find a unique solution to these equations in the interior of X, namely the point $A = (\frac{1}{2}, \frac{1}{3})$, so this must be the unique maximum of f. (If we had found more than one critical point, then it would have been necessary to calculate the value of f at each one of them and then choose the point(s) giving the maximum value.)

Exercise 2.6. (An extension of Rolle's Theorem) Let E be a normed vector space, X a compact subset, whose interior is not empty, and f a real-valued continuous function defined on X, which is differentiable on the interior of X. Show that if f is constant on the boundary of X, then f has a critical point in the interior of X.

2.6   Differentiability of the Norm

If E is a normed vector space with norm $\|\cdot\|$, then $\|\cdot\|$ is itself a mapping from E into $\mathbb{R}$ and we may consider its differentiability. We will write $\|x\|'$ for the differential of the norm at x (if it exists). We should first notice that the norm is never differentiable at the origin. Suppose that $\|0\|'$ exists. Then for small nonzero values of h, we have

$$\|h\| = \|0\|'h + o(h) \implies \lim_{h \to 0} \left(1 - \|0\|' \frac{h}{\|h\|} \right) = 0$$

and

$$\|h\| = \|-h\| = -\|0\|'h + o(h) \implies \lim_{h \to 0} \left(1 + \|0\|' \frac{h}{\|h\|} \right) = 0.$$

Summing the two limits we obtain $2 = 0$, which is clearly a contradiction. Hence $\|0\|'$ does not exist.

The norm may or may not be differentiable at a point other than the origin. For example, the norm $\|\cdot\|_p$ defined on $\mathbb{R}^n$ is differentiable at all points other than the origin if $p \in (1, \infty)$; however, this is not the case if $p = 1$ or $p = \infty$. We will study these norms more in detail at the end of the section. At points where the differential exists, we have the following interesting result:

Proposition 2.11. *Let E be a normed vector space and $\|\cdot\|$ its norm. If $\|\cdot\|$ is differentiable at $a \neq 0$ and $\lambda > 0$, then $\|\cdot\|$ is differentiable at λa and $\|\lambda a\|' = \|a\|'$. In addition, $|\|a\|'|_{E^*} = 1$.*

Proof. If $\|\cdot\|$ is differentiable at a, $\lambda > 0$ and $h \in E \setminus \{0\}$ small, then we have

$$\|\lambda a + h\| = \lambda \|a + \frac{h}{\lambda}\|$$

$$= \lambda \left(\|a\| + \|a\|' \frac{h}{\lambda} + o\left(\frac{h}{\lambda} \right) \right)$$

$$= \|\lambda a\| + \|a\|'h + o(h).$$

It follows that $\|\lambda a\|'$ exists and $\|\lambda a\|' = \|a\|'$.

Now let us show that $|\|a\|'|_{E^*} = 1$. Consider the function

$$f : \mathbb{R}_+^* \longrightarrow \mathbb{R}, \lambda \longmapsto \|\lambda a\|.$$

For a given $\lambda \in \mathbb{R}_+^*$ and $h \in \mathbb{R}$ sufficiently small, we have

$$\|(\lambda + h)a\| = (\lambda + h)\|a\|$$

and so

$$\lim_{h \to 0} \frac{\|(\lambda + h)a\| - \|\lambda a\|}{h} = \lim_{h \to 0} \frac{h\|a\|}{h} = \|a\|.$$

Therefore $\dot{f}(\lambda) = \|a\|$ for all values of λ. On the other hand, $f = \|\cdot\| \circ \phi$, where $\phi(\lambda) = \lambda a$, and so

$$f'(\lambda)s = \|\lambda a\|' sa = s\|a\|'a.$$

This implies that $\dot{f}(\lambda) = \|a\|'a$ and hence $\|a\|'a = \|a\|$. It follows that $\|\|a\|'\|_{E^*} \geq 1$.

Now let us show that $\|\|a\|'\|_{E^*} \leq 1$. As $\|\cdot\|$ is differentiable at a, we may write

$$\|a + \lambda x\| - \|a\| - \|a\|' \lambda x = \|\lambda x\| \epsilon(\lambda x),$$

where $\lim_{h \to 0} \epsilon(h) = 0$. This implies that

$$\|a\|' \lambda x| \leq |\|a + \lambda x\| - \|a\|| + |\|\lambda x\| \epsilon(\lambda x)|,$$

hence

$$\|a\|' x| \leq \|x\|(1 + |\epsilon(\lambda x)|).$$

Letting λ go to 0 we obtain

$$\|a\|' x| \leq \|x\|,$$

which implies that $\|\|a\|'\|_{E^*} \leq 1$. This finishes the proof. □

Corollary 2.6. *The norm is differentiable on $E \setminus \{0\}$ if and only if it is differentiable on the unit sphere.*

We have already briefly spoken of the differentiability of the norms $\|\cdot\|_p$ defined on $\mathbb{R}^n$. We will now consider these norms in more detail. We will use the notation $\partial_i \|x\|_p$ for the ith partial derivative at x, which is simpler than $\frac{\partial \|\cdot\|_p}{\partial x_i}(x)$. There are three cases to consider, namely $\|\cdot\|_p$, with $1 < p < \infty$, $\|\cdot\|_1$ and $\|\cdot\|_\infty$.

Case 1. $\|\cdot\|_p$, $1 < p < \infty$: all partial derivatives exist and are continuous on the open set $\mathbb{R}^n \setminus \{0\}$.
If $x_i > 0$, then

$$\partial_i \|x\|_p = \frac{1}{p}(|x_1|^p + \cdots + |x_n|^p)^{\frac{1}{p}-1} p x_i^{p-1} = \|x\|_p^{1-p} x_i^{p-1}$$

and, for $x_i < 0$, we have

$$\partial_i \|x\|_p = -\frac{1}{p}(|x_1|^p + \cdots + |x_n|^p)^{\frac{1}{p}-1} p(-x_i)^{p-1} = \|x\|_p^{1-p}(-x_i)^{p-1}.$$

The case where $x_i = 0$ is a little more delicate. We are interested in the following limit (if it exists):

$$\lim_{t \to 0} \frac{1}{t}\left(\|x + te_i\|_p - \|x\|_p\right) = \lim_{t \to 0} \frac{1}{t}\left(\left(|t|^p + \|x\|_p^p\right)^{\frac{1}{p}} - \|x\|_p\right),$$

where e_i is the ith element of the standard basis of $\mathbb{R}^n$. If we divide the numerator of the expression on the right-hand side by $\|x\|_p$ we obtain

$$\left(\left(\frac{|t|}{\|x\|_p}\right)^p + 1\right)^{\frac{1}{p}} - 1 = \left(1 + \frac{1}{p}\left(\frac{|t|}{\|x\|_p}\right)^p + o\left(\left(\frac{|t|}{\|x\|_p}\right)^p\right)\right) - 1$$

$$= \frac{1}{p}\left(\frac{|t|}{\|x\|_p}\right)^p + o(|t|^p).$$

As

$$\lim_{t \to 0} \frac{\|x\|_p}{t}\left(\frac{1}{p}\left(\frac{|t|}{\|x\|_p}\right)^p + o(|t|^p)\right) = 0,$$

we have

$$\partial_i \|x\|_p = 0.$$

Hence the partial derivatives are defined and continuous on $\mathbb{R}^n \setminus \{0\}$ and so the norm $\|\cdot\|_p$ is of class C^1 on this set.

Case 2. $\|\cdot\|_1$: all partial derivatives are defined and continuous on the open set $S = \{x \in \mathbb{R}^n : x_i \neq 0 \text{ for all } i\}$.
We have

$$\|x\|_1 = |x_1| + \cdots + |x_n|.$$

If $x_i = 0$, then

$$\frac{1}{t}(\|x + te_i\|_1 - \|x\|_1) = \frac{|t|}{t}$$

and so $\partial_i \|x\|_1$ does not exist, which implies that the differential is not defined at a point x with a coordinate whose value is 0. Now suppose that this not the case and that $t \in \mathbb{R}$ is such that $|t| < |x_i|$. For $x_i > 0$

$$\frac{1}{t}(\|x + te_i\|_1 - \|x\|_1) = 1$$

and for $x_i < 0$

$$\frac{1}{t}(\|x + te_i\|_1 - \|x\|_1) = -1.$$

It follows that in the first case $\partial_i \|x\|_1 = 1$ and in the second $\partial_i \|x\|_1 = -1$. If all the coordinates of a point x are nonzero, then we may find a neighbourhood N of x such that, if $y \in N$, then each coordinate y_i of y is nonzero and has the same sign as that of x_i. Hence all partial derivatives of the norm are defined and continuous on N. It follows that $\|\cdot\|_1$ is of class C^1 on S.

Case 3. $\|\cdot\|_\infty$: all partial derivatives are defined and continuous on the open set $T = \{x \in \mathbb{R}^n : |x_i| = \|x\|_\infty \text{ for a unique } i\}$.

Let $x \neq 0$ with $|x_k| = \|x\|_\infty$. First let us suppose that x_k is not unique, i.e., there is an index l, with $l \neq k$, such that $|x_l| = |x_k|$. We now take $h_k, h_l \in \mathbb{R}^*$ with the same absolute value and such that h_k has the same sign as x_k and h_l the sign opposite to that of x_l. We set $h = (h_1, \ldots, h_n)$, where $h_i = 0$ for $i \neq k, l$. Then we have

$$\|x + h\|_\infty = |x_k + h_k| = |x_k| + |h_k| = \|x\|_\infty + \|h\|_\infty$$

and

$$\|x - h\|_\infty = |x_l - h_l| = |x_l| + |h_l| = \|x\|_\infty + \|h\|_\infty.$$

By addition we obtain

$$\|x + h\|_\infty + \|x - h\|_\infty - 2\|x\|_\infty = 2\|h\|_\infty. \tag{2.1}$$

If we suppose that the differential exists at x, then we have

$$\|x + h\|_\infty - \|x\|_\infty - \|x\|_\infty' h = o(h)$$

and

$$\|x - h\|_\infty - \|x\|_\infty + \|x\|_\infty' h = o(h).$$

An addition of the two expressions gives us

$$\|x + h\|_\infty + \|x - h\|_\infty - 2\|x\|_\infty = o(h). \tag{2.2}$$

Now, from the (2.1) and (2.2) we obtain

$$2\|h\|_\infty = o(h) \implies \lim_{h \to 0} \frac{\|h\|_\infty}{\|h\|_\infty} = 0,$$

which is clearly false. It follows that the norm is not differentiable at x.

Now let us suppose that x_k is unique. Let e_i be the ith member of the standard basis of $\mathbb{R}^n$ and $t \in \mathbb{R}^*$ small. Then

$$\|x + te_i\|_\infty = \begin{cases} |x_k| & \text{for } i \neq k \\ |x_k + t| & \text{for } i = k \end{cases}.$$

Therefore

$$\partial_i \|x\|_\infty = \begin{cases} 0 & \text{if } i \neq k \\ 1 & \text{if } i = k \text{ and } x_k > 0 \\ -1 & \text{if } i = k \text{ and } x_k < 0 \end{cases}.$$

Also, there is a neighbourhood N of x such that, if $y \in N$, then y_k is nonzero, has the same sign as x_k and $\|y\|_\infty = |y_k|$ for a unique k. It follows that $\partial_i \|y\|_\infty = \partial_i \|x\|_\infty$ for all i. Thus the partial derivatives are defined and continuous on N and so the norm is of class C^1 on T.

Remark. We have seen above that certain norms on $\mathbb{R}^n$ are differentiable at all points other than the origin and others not. Thus equivalent norms on a vector space may have different differentiability properties.

Appendix: Gâteaux Differentiability

There is another differential which is often used. We have already defined the directional derivative at a point with respect to a nonzero vector for a real-valued function. In fact, we can extend this definition to mappings whose image lies in any normed vector space. Let O be an open subset of a normed vector space E, f a mapping defined on O whose image lies in a normed vector space F, $a \in O$ and u a nonzero element of E. If the limit

$$\lim_{t \longrightarrow 0} \frac{f(a + tu) - f(a)}{t}$$

exists, then we note this derivative $\partial_u f(a)$. It is called the *directional derivative of f at a in the direction u*. If the directional derivative is defined in all directions and there is a continuous linear mapping ϕ from E into F such that for all $u \in E$

$$\partial_u f(a) = \phi(u),$$

then we say that f is Gâteaux-differentiable at a and that ϕ is the *Gâteaux differential* of f at a. If a mapping f is differentiable at a point a, then clearly all its directional derivatives exist and we have $\partial_u f(a) = f'(a)u$. Thus f is Gâteaux-differentiable at a. However, the Gâteaux differential may exist without the differential existing. Here is an example. If the mapping $f : \mathbb{R}^2 \longrightarrow \mathbb{R}$ is defined by

$$f(x, y) = \begin{cases} \frac{x^6}{x^8 + (y - x^2)^2} & \text{if } (x, y) \neq (0, 0) \\ 0 & \text{if } (x, y) = (0, 0) \end{cases},$$

then $\partial_{(u,v)} f(0, 0) = 0$ for all $(u, v) \in \mathbb{R}^2$ and so the Gâteaux differential exists at the origin. However, $f(x, x^2) = x^{-2}$ and so f is not continuous at the origin and *a fortiori* not differentiable.

Another point should also be made, namely that the existence of directional derivatives at a point does not imply that the mapping is Gâteaux differentiable. Let us consider the mapping $g : \mathbb{R}^2 \longrightarrow \mathbb{R}$ defined by

$$g(x, y) = \begin{cases} \frac{x(x^2 - 3y^2)}{x^2 + y^2} & \text{if } (x, y) \neq (0, 0) \\ 0 & \text{if } (x, y) = (0, 0) \end{cases}.$$

Then $\partial_{(u,v)} g(0, 0) = g(u, v)$ for all $(u, v) \in \mathbb{R}^2$. As g is not linear, g is not Gâteaux differentiable at $(0, 0)$.

To distinguish the differential from the Gâteaux differential, the differential is often referred to as the Fréchet differential. From what we have seen, we have the implications:

Fréchet differentiable $\implies$ Gâteaux differentiable $\implies$ existence of
directional derivatives

and the implications are not reversible.

In fact, the Gâteaux differentiability of the norm is closely related to the geometry
of the unit sphere. The book by Beauzamy [3] handles this subject in some detail.

Chapter 3
Mean Value Theorems

In elementary calculus we learn the mean value theorem:

Theorem 3.1. *Let f be a real-valued function defined on a closed bounded interval $[a,b] \subset \mathbb{R}$. If f is continuous on $[a,b]$ and differentiable on (a,b), then there is a point $c \in (a,b)$ such that*

$$f(b) - f(a) = \dot{f}(c)(b-a).$$

In this chapter we introduce some generalizations of this mean value theorem, which are also often called mean value theorems. These theorems have many applications, some of which we will introduce here.

We will need a generalization of the usual derivative of a real-valued function. If $\lambda \in \mathbb{R}^*$ and $x \in E$, a normed vector space, we will often write $\frac{x}{\lambda}$ for $\frac{1}{\lambda}x$, i.e., the scalar product of $\frac{1}{\lambda}$ with x. Suppose that $I \subset \mathbb{R}$ is an interval and that f is a mapping whose image lies in a normed vector space E. Then we can define a derivative in the usual way. If $a \in I$ and the limit

$$\lim_{t \to 0} \frac{f(a+t) - f(a)}{t}$$

exists, then we call this limit, which we write $\frac{df}{dx}(a)$ or $\dot{f}(a)$, the *derivative of f* at a. If the interval I is open, then we can prove, as in Proposition 2.6, that f is differentiable at $a \in I$ if and only if f has a derivative at a. In this case we have

$$f'(a)t = t\dot{f}(a),$$

If the function f is differentiable at all points $x \in I$, then we have a function $\dot{f} : I \longrightarrow E$ defined in a natural way. We should also notice that the norm of $f'(x)$ is equal to that of $\dot{f}(x)$ and so the continuity of $\dot{f}$ at x is equivalent to that of f' at x. Therefore f is of class C^1, if $\dot{f}$ is continuous.

R. Coleman, *Calculus on Normed Vector Spaces*, Universitext,
DOI 10.1007/978-1-4614-3894-6_3, © Springer Science+Business Media New York 2012

Remark. We may extend the term "differentiable" to a mapping defined on any interval I. If $f : I \longrightarrow E$ has a derivative $\dot{f}(x)$ at every point $x \in I$, then we say that f is differentiable. In this case if the mapping $\dot{f}$ is continuous, then we say that f is of class C^1.

3.1 Generalizing the Mean Value Theorem

Our first result is very close to the classical mean value theorem.

Theorem 3.2. *Let O be an open subset of a normed vector space E and a and b elements of E with $[a,b] \subset O$. If $f : O \longrightarrow \mathbb{R}$ is differentiable, then there is an element $c \in (a,b)$ such that*

$$f(b) - f(a) = f'(c)(b-a).$$

Proof. Let $u : [0,1] \longrightarrow E$ be the mapping defined by $u(t) = a + t(b-a)$. u is continuous on $[0,1]$ and differentiable on $(0,1)$. It follows that the real-valued function $g = f \circ u$ is continuous on $[0,1]$ and differentiable on $(0,1)$. For $t \in (0,1)$ we have

$$\dot{g}(t) = g'(t)1 = f'(u(t)) \circ u'(t)1 = f'(u(t))(b-a).$$

From Theorem 3.1 there exists $\theta \in (0,1)$ such that

$$g(1) - g(0) = \dot{g}(\theta)(1-0) = \dot{g}(\theta).$$

This can be written

$$f(b) - f(a) = f'(u(\theta))(b-a)$$

and hence the result.   □

Remark. If $E = \mathbb{R}^n$, then this result can be written

$$f(b) - f(a) = \sum_{i=1}^{n} \left(\frac{\partial f}{\partial x_i}(c) \right) (b_i - a_i).$$

Let S be a subset of a normed vector space. If $A, B \subset S$ are nonempty and form a partition of S and there exist open subsets $U, V \subset E$ such that $A = S \cap U$ and $B = S \cap V$, then we say that the pair (A, B) is a *disconnection* of S. If S has a disconnection, then we say that S is *disconnected*, otherwise we say that S is *connected*. Clearly, if S is open, then S is disconnected if and only if it is the disjoint union of two nonempty open sets.

Exercise 3.1. Show that, if E is a normed vector space and S a convex subset of E, then S is connected. (This exercise generalizes Proposition 1.10.)

If S is a subset of a normed vector space E and γ a continuous mapping from $[0, 1]$ into S, then we say that γ is a *path* in S. If there is a partition $P : 0 = t_0 < t_1 < \cdots < t_p = 1$ of $[0, 1]$ such that γ restricted to each subinterval $[t_i, t_{i+1}]$ is affine, then we say that γ is *polygonal*. In this case the image of γ restricted to $[t_i, t_{i+1}]$ is the segment $[\gamma(t_i), \gamma(t_{i+1})]$.

Lemma 3.1. *If O is an open connected subset of a normed vector space E and $a, b \in O$, then there is a polygonal path lying in O joining a to b.*

Proof. We define a relation "$\sim$" on O by writing $x \sim y$ if there is a polygonal path joining x to y. There is no difficulty in seeing that this is an equivalence relation. It is sufficient to show that there is a unique equivalence class. To do so we take $a \in O$ and let $[a]$ be its equivalence class. Let $x \in [a]$. As O is open, there is an open ball $B(x, r)$ included in O. If $y \in B(x, r)$, then the path $\gamma : [0, 1] \longrightarrow E, t \longmapsto x + t(y - x)$ is affine, has its image in $B(x, r)$ and connects y to x. It follows that $y \in [a]$. Hence $B(x, r) \subset [a]$ and so $[a]$ is an open set. Therefore, if there is more than one equivalence class, O has a disconnection, namely a given equivalence class and the union of the others. This is a contradiction and the result follows.   □

We are now in a position to generalize a result from elementary analysis, namely, if the derivative of a real-valued differentiable function defined on an open interval of $\mathbb{R}$ vanishes, then the function is constant.

Theorem 3.3. *If O is an open connected subset of a normed vector space E, $f : O \longrightarrow \mathbb{R}$ a differentiable function and the differential f' vanishes, then f is constant.*

Proof. Fix $a \in O$ and let x be any other element of O. From the lemma there is a polygonal path γ connecting a to x. From Theorem 3.2 and using the notation above, we have

$$f(\gamma(t_{i+1})) - f(\gamma(t_i)) = 0$$

for $i = 0, \ldots, p - 1$ and so $f(x) = f(a)$.   □

Let us consider the mapping

$$f : \mathbb{R} \longrightarrow \mathbb{R}^2, t \longmapsto (\cos(2\pi t), \sin(2\pi t)).$$

Here $f(1) - f(0) = 0$. However, $f'(t) = (-2\pi \sin(\pi t)\mathrm{d}t, 2\pi \cos(\pi t)\mathrm{d}t)$, where $\mathrm{d}t$ is the identity on $\mathbb{R}$, and so we cannot find $t_0 \in (0, 1)$ such that $f(1) - f(0) = f'(t_0)(1 - 0)$. Therefore we cannot generalize Theorem 3.2 to mappings whose image lies in a general normed vector space. However, a consequence of this theorem is:

$$|f(b) - f(a)| \leq \sup_{z \in (a,b)} |f'(z)|_{E^*}(b - a).$$

This can be generalized as we will soon see. We will proceed by steps.

Theorem 3.4. *Let* $[a, b]$ *be an interval of* $\mathbb{R}$, F *a normed vector space and* $f :$ $[a, b] \longrightarrow F$ *and* $g : [a, b] \longrightarrow \mathbb{R}$ *both continuous on* $[a, b]$ *and differentiable on* (a, b). *If* $\|\dot{f}(t)\|_F \leq \dot{g}(t)$ *for all* $t \in (a, b)$, *then* $\|f(b) - f(a)\|_F \leq g(b) - g(a)$.

Proof. Let us take $\epsilon > 0$. We define $A(\epsilon)$ to be the set of elements $x \in [a, b]$ such that for all $y \in [a, x]$ we have

$$\|f(y) - f(a)\|_F \leq g(y) - g(a) + \epsilon(y - a).$$

Clearly $a \in A(\epsilon)$. Also, if $x \in A(\epsilon)$ and $\tilde{x} \in [a, x]$, then $\tilde{x} \in A(\epsilon)$, and so $A(\epsilon)$ is an interval. Let c be the least upper bound of $A(\epsilon)$. We claim that $c \in A(\epsilon)$. If $c = a$, then there is nothing to prove, so suppose that this not the case. If $x \in (a, c)$, then

$$\|f(x) - f(a)\|_F \leq g(x) - g(a) + \epsilon(x - a).$$

Using the continuity of f and g, we see that this inequality applies for $x = c$ and therefore $c \in A(\epsilon)$.

We will now show that $c = b$. If $c < b$, then, given the hypothesis on the derivatives, we may find $\eta \in (0, b - c)$ such that

$$\left\| \frac{f(x) - f(c)}{x - c} \right\|_F \leq \frac{g(x) - g(c)}{x - c} + \epsilon$$

for $x \in (c, c + \eta)$, from which we obtain

$$\|f(x) - f(c)\|_F \leq g(x) - g(c) + \epsilon(x - c).$$

Now, using the triangle inequality, we have

$$
\begin{aligned}
\|f(x) - f(a)\|_F &\leq \|f(x) - f(c)\|_F + \|f(c) - f(a)\|_F \\
&\leq \big(g(x) - g(c) + \epsilon(x - c)\big) + \big(g(c) - g(a) + \epsilon(c - a)\big) \\
&= g(x) - g(a) + \epsilon(x - a),
\end{aligned}
$$

which implies that $(c, c + \eta) \subset A(\epsilon)$. This contradicts the definition of c. It follows that $c = b$ and we may write

$$\|f(b) - f(a)\|_F \leq g(b) - g(a) + \epsilon(b - a).$$

Letting ϵ converge to 0 we obtain the result.   □

Corollary 3.1. *Let* $a, b \in \mathbb{R}$, *with* $a < b$, F *a normed vector space and* $f :$ $[a, b] \longrightarrow F$ *continuous on* $[a, b]$ *and differentiable on* (a, b). *If there is a constant* K *such that* $|\dot{f}(t)|_F \leq K$, *then*

$$\|f(b) - f(a)\|_F \leq K(b - a).$$

Proof. To establish this result, it is sufficient to set $g(t) = Kt$ and apply the theorem. □

The next result, often called the mean value inequality, is the generalization referred to above.

Corollary 3.2. *Let E and F be normed vector spaces, O an open subset of E and $f : O \longrightarrow F$ differentiable on O. If the segment $[a, b] \subset O$, then*

$$\|f(b) - f(a)\|_F \leq \sup_{x \in (a,b)} |f'(x)|_{\mathcal{L}(E,F)} \|b - a\|_E.$$

Proof. If $\sup_{x \in (a,b)} |f'(x)|_{\mathcal{L}(E,F)} = \infty$, then there is nothing to prove, so suppose that this is not the case. Let $u : [0, 1] \longrightarrow E$ be defined by $u(t) = (1 - t)a + tb$. If $g = f \circ u$, then g is continuous on $[0, 1]$ and differentiable on $(0, 1)$, with

$$\dot{g}(t) = f'(u(t)) \circ u'(t)1 = f'(u(t))(b - a).$$

Therefore

$$\|\dot{g}(t)\|_F \leq \sup_{x \in (a,b)} |f'(x)|_{\mathcal{L}(E,F)} \|b - a\|_E.$$

From Corollary 3.1 we have

$$\|g(1) - g(0)\|_F \leq \sup_{x \in (a,b)} |f'(x)|_{\mathcal{L}(E,F)} \|b - a\|_E (1 - 0),$$

i.e.,

$$\|f(b) - f(a)\|_F \leq \sup_{x \in (a,b)} |f'(x)|_{\mathcal{L}(E,F)} \|b - a\|_E.$$

This is what we set out to prove. □

Exercise 3.2. Let E be a normed vector space and f a differentiable mapping from E into itself. Suppose that there exists $k \in (0, 1)$ such that $|f'(x) - \mathrm{id}_E|_{\mathcal{L}(E)} \leq k$ for all $x \in E$. Prove that f is injective and that the inverse image of a bounded set is bounded.

Exercise 3.3. Let E and F be normed vector spaces, O an open subset of E, a an element of O and f a continuous mapping from O into F, which is differentiable on $O \setminus \{a\}$. Suppose now that there is a continuous linear mapping l from E into F such that

$$\lim_{x \to a} |f'(x) - l|_{\mathcal{L}(E,F)} = 0.$$

Show that f is differentiable at a, with $f'(a) = l$.

Corollary 3.2 can be used to prove another useful result.

Corollary 3.3. *Let E and F be normed vector spaces, O an open subset of E and $f : O \longrightarrow F$ differentiable. If the segment $[a, b]$ lies in O, then we have*

$$\| f(b) - f(a) - f'(a)(b-a) \|_F \leq \sup_{x \in (a,b)} |f'(x) - f'(a)|_{\mathcal{L}(E,F)} \|b - a\|_E.$$

Proof. If we set $\phi(x) := f(x) - f'(a)x$, then ϕ is differentiable and $\phi'(x) = (f'(x) - f'(a))$. Applying the previous corollary we obtain the result. $\square$

Using Corollary 3.2 again, we can generalize Theorem 3.3. The proof is analogous to that of Theorem 3.3.

Theorem 3.5. *Let E and F be normed vector spaces, O an open connected subset of E and $f : O \longrightarrow F$ a differentiable mapping. If f' vanishes, then f is constant.*

Exercise 3.4. Let E and F be normed vector spaces, O an open connected subset of E and f a differentiable mapping from O into F with constant differential $\alpha \in \mathcal{L}(E, F)$. Show that there is a constant $c \in F$ such that $f(x) = \alpha x + c$.

3.2 Partial Differentials

In this section we will generalize the notion of partial derivative and then extend certain results concerning partial derivatives.

Let $E_1, \ldots, E_n$ and F be normed vector spaces. We set

$$E = E_1 \times \cdots \times E_n$$

and define a norm on E as usual by

$$\|(x_1, \ldots, x_n)\|_E = \max_k \|x_k\|_{E_k}.$$

Now let O be an open subset of E and f a mapping from O into F. If we take a point $a \in O$ and let the kth coordinate vary and fix the others, then we obtain a mapping $f_{a,k}$ from E_k into F, defined on an open subset of E_k containing a_k. If $f_{a,k}$ is differentiable at a_k, then we call the differential $f'_{a,k}(a_k) \in \mathcal{L}(E_k, F)$ the *kth partial differential of f at a* and write $\partial_k f(a)$ for $f'_{a,k}(a_k)$. In the case where $E = \mathbb{R}^n$, f has a kth partial differential at a if and only if f has a kth partial derivative at a.

Exercise 3.5. Why is the mapping $f_{a,k}$ defined on an open subset of E_k?

We now extend certain results found in Chap. 2. We begin with Proposition 2.5.

Proposition 3.1. *If $f : O \longrightarrow F$ is differentiable at $a \in O$, then all the partial differentials $\partial_k f(a)$ exist and*

$$f'(a)h = \sum_{k=1}^{n} \partial_k f(a)h_i,$$

where $h = (h_1, \ldots, h_n) \in E$.

Proof. For $k = 1, \ldots, n$, we define the mapping $i_{a,k}$ from E_k into E by

$$i_{a,k}(t) = (a_1, \ldots, t, \ldots, a_n),$$

where t is in the kth position. $i_{a,k}$ is an affine mapping and so differentiable. Also, $f_{a,k} = f \circ i_{a,k}$. Therefore, if f is differentiable at a, then the partial derivative $\partial_k f(a)$ is defined and

$$\partial_k f(a)h_k = f'(a)(i'_k(a_k)h_k) = f'(a)(0, \ldots, h_k, \ldots, 0).$$

Using the linearity of $f'(x)$, we obtain

$$f'(a)h = \sum_{k=1}^{n} \partial_k f(a)h_k.$$

This finishes the proof.   □

If $f : O \longrightarrow F$ is differentiable, then for each k we have a mapping $\partial_k f$ from O into $\mathcal{L}(E_k, F)$, which is called a *partial differential (mapping)*. If f is of class C^1 and $a \in O$, then

$$\|(\partial_k f(a+x) - \partial_k f(a))h_k\|_F = \|(f'(a+x) - f'(a))(0, \ldots, h_k, \ldots, 0)\|_E$$
$$\leq |f'(a+x) - f'(a)|_{\mathcal{L}(E,F)}\|h_k\|_{E_k},$$

which implies that

$$|\partial_k f(a+x) - \partial_k f(a)|_{\mathcal{L}(E_k,F)} \leq |f'(a+x) - f'(a)|_{\mathcal{L}(E,F)}.$$

It follows that the partial differential $\partial_k f$ is continuous at a. To sum up, we have proved the

Proposition 3.2. *If f is of class C^1, then its partial differentials exist and are continuous.*

The converse of this proposition is also true, namely, if f has continuous partial differentials, then f is of class C^1. We will prove a preliminary result, which generalizes Theorem 2.2, and the converse will follow directly. To do so, we will employ a mean value theorem, namely Corollary 3.2.

Theorem 3.6. *Let $E_1, \ldots, E_n$ and F be normed vector spaces, O an open subset of $E = E_1 \times \cdots \times E_n$ and $a \in O$. If f is a mapping from O into F having*

continuous partial differentials on a neighbourhood V of a, then f is continuously differentiable at a.

Proof. Let $h = (h_1, \ldots, h_n) \in E$. We set

$$h^0 = (0, \ldots, 0) \qquad \text{and} \qquad h^i = (h_1, \ldots, h_i, 0, \ldots, 0)$$

for $i = 1, \ldots, n$. Clearly $h^n = h$. Let $\bar{\delta} > 0$ be such that the open ball $B(x, \bar{\delta}) \subset V$. If $\|h\| < \bar{\delta}$, then $a + h^i \in B(a, \bar{\delta})$ for all i. For $x \in B(a, \bar{\delta})$ we set

$$g(x) = f(x) - f(a) - \sum_{i=1}^{n} \partial_i f(a)(x_i - a_i).$$

Then $g(a) = 0$ and all the partial differentials of g are defined on V:

$$\partial_i g(x) = \partial_i f(x) - \partial_i f(a).$$

Let us fix $\epsilon > 0$. As the partial differentials $\partial_i f$ are defined and continuous on V, there is a $\delta > 0$, with $\delta \leq \bar{\delta}$, such that

$$\|h\| < \delta \implies |\partial_i f(a+h) - \partial_i f(a)|_{\mathcal{L}(E_i, F)} < \epsilon$$

for $i = 1, \ldots, n$. Now, using Corollary 3.2, we have

$$\|g(a+h)\|_F \leq \sum_{i=1}^{n} \|g(a+h^i) - g(a+h^{i-1})\|_F \leq \sum_{i=1}^{n} \epsilon \|h_i\|_{E_i} \leq n\epsilon \|h\|_E,$$

i.e.,

$$\left\| f(a+h) - f(a) - \sum_{i=1}^{n} \partial_i f(a) h_i \right\|_F \leq n\epsilon \|h\|_E.$$

It follows that f is differentiable at a and

$$f'(a)h = \sum_{i=1}^{n} \partial_i f(a) h_i.$$

We also have

$$\|(f'(a+k) - f'(a))h\|_F \leq \sum_{i=1}^{n} |\partial_i f(a+k) - \delta_i f(a)|_{\mathcal{L}(E_i, F)} \|h_i\|_{E_i}$$

$$\leq \sum_{i=1}^{n} |\partial_i f(a+k) - \partial_i f(a)|_{\mathcal{L}(E_i, F)} \|h\|_E,$$

which implies that

$$|f'(a+k) - f'(a)|_{\mathcal{L}(E,F)} \leq \sum_{i=1}^{n} |\partial_i f(a+k) - \partial_i f(a)|_{\mathcal{L}(E_i,F)}$$

and so f' is continuous at a. $\square$

Corollary 3.4. *If $E_1, \ldots, E_n$ and F are normed vector spaces, O an open subset of $E = E_1 \times \cdots \times E_n$ and f a mapping from O into F having continuous partial differentials on O, then f is of class C^1.*

Putting our work in this section together, we obtain the following generalization of Theorem 2.3:

Theorem 3.7. *If $E_1, \ldots, E_n$ and F are normed vector spaces, O an open subset of $E = E_1 \times \cdots \times E_n$ and f a mapping from O into F, then f is of class C^1 if and only if f has continuous partial differentials defined on O.*

3.3 Integration

In elementary calculus courses one learns the following result:

Theorem 3.8. *Let I and $[a,b]$ be intervals of $\mathbb{R}$, with $[a,b]$ closed and bounded, and f a continuous real-valued function defined on $[a,b] \times I$. Then the function*

$$g : I \longrightarrow \mathbb{R}, x \longmapsto \int_a^b f(t,x)\mathrm{d}t$$

is continuous. If, in addition, the partial derivative $\frac{\partial f}{\partial x}$ is defined everywhere on I and is continuous, then g has a derivative at all points of I and

$$\frac{\mathrm{d}g}{\mathrm{d}x}(x) = \int_a^b \frac{\partial f}{\partial x}(t,x)\mathrm{d}t.$$

We will now generalize this result. However, before doing so, we need to define the integral of a mapping whose image lies in a Banach space. This we will do in this section and in the next we will handle the generalization.

We first recall the definition of the Riemann integral. Let $[a,b]$ be a closed bounded interval of $\mathbb{R}$ and f a real-valued bounded function defined on $[a,b]$. For a partition $P : a = x_0 < x_1 < \cdots < x_n = b$ and $i = 1, \ldots, n$, we define

$$m_i = \inf_{x \in [x_{i-1}, x_i]} f(x) \quad \text{and} \quad M_i = \sup_{x \in [x_{i-1}, x_i]} f(x)$$

and then

$$L(P, f) = \sum_{i=1}^{n} m_i (x_i - x_{i-1}) \quad \text{and} \quad U(P, f) = \sum_{i=1}^{n} M_i (x_i - x_{i-1}).$$

Now we set

$$\mathcal{L} \int_a^b f = \sup L(P, f) \quad \text{and} \quad \mathcal{U} \int_a^b f = \inf U(P, f),$$

where the sup and inf are taken over all partitions of $[a, b]$. These expressions are called respectively the lower and upper integrals of f. If these two integrals have the same value, then we say that f is (Riemann) integrable and we write

$$\int_a^b f \quad \text{or} \quad \int_a^b f(x) \mathrm{d}x$$

for the common value of the lower and upper integrals. This common value is called the (Riemann) integral of f.

It is not possible to generalize directly this definition of integral if we replace the image space $\mathbb{R}$ by a general normed vector space E, because there is no notion of inf or sup in such a space. However, if (f_n) is a sequence of real-valued integrable functions converging uniformly to a real-valued function f, then f is integrable and the sequence $(\int_a^b f_n)$ converges to $\int_a^b f$. It is this property which allows us to generalize the integral.

Let $[a, b]$ be a closed bounded interval of $\mathbb{R}$ and E a Banach space. We define a norm $\| \cdot \|$ on $\mathcal{B}([a, b], E)$, the vector space of bounded mappings defined on $[a, b]$ with image in E, by setting

$$\|f\| = \sup_{x \in [a,b]} \|f(x)\|_E.$$

The normed vector space $\mathcal{B}([a, b], E)$ is a Banach space. We say that $f : [a, b] \longrightarrow E$ is a *step mapping* if there is a partition $P : a = x_0 < x_1 < \cdots < x_n = b$ of $[a, b]$ and elements $c_1, \ldots, c_n \in E$ such that $f(x) = c_i$ on the interval (x_{i-1}, x_i). The set of step mappings, which we denote $\mathcal{S}([a, b], E)$, forms a vector subspace of $\mathcal{B}([a, b], E)$. The closure of $\mathcal{S}([a, b], E)$, for which we will write $\mathcal{R}([a, b], E)$, is also a vector subspace of $\mathcal{B}([a, b], E)$ and its elements are called *regulated mappings*. We can characterize regulated mappings in the following useful way.

Proposition 3.3. *An element of $\mathcal{B}([a, b], E)$ is regulated if and only if it has a left limit at any point $x \in (a, b]$ and a right limit at any point $x \in [a, b)$.*

Proof. Let f be a regulated mapping and (f_n) a sequence of step mappings converging to f. Take $\epsilon > 0$. There exists an f_n such that

$$\|f(t) - f_n(t)\|_E < \frac{\epsilon}{2}$$

for all $t \in [a, b]$. We now fix $x \in (a, b]$. For η sufficiently small, f_n is constant on the interval $(x - \eta, x)$ and so, for $s, t \in (x - \eta, x)$, we have

$$\|f(s) - f(t)\|_E \le \|f(s) - f_n(s)\|_E + \|f_n(s) - f_n(t)\|_E$$
$$+ \|f_n(t) - f(t)\|_E < \frac{\epsilon}{2} + 0 + \frac{\epsilon}{2} = \epsilon.$$

It follows, using the fact that E is a Banach space, that $\lim_{t \to x-} f(t)$ exists. A similar argument shows that $\lim_{t \to x+} f(t)$ exists if $x \in [a, b)$.

Now suppose that the mapping f satisfies the condition on left and right limits and let us take $\epsilon > 0$. For each point $x \in (a, b]$ there exists $c_x < x$ such that

$$s, t \in (c_x, x) \implies \|f(s) - f(t)\|_E < \epsilon,$$

and for each point $x \in [a, b)$ there exists $d_x > x$ such that

$$s, t \in (x, d_x) \implies \|f(s) - f(t)\|_E < \epsilon.$$

Using the compacity of $[a, b]$, we see that there are elements $x_1, \ldots, x_n \in [a, b]$ such that

$$[a, b] \subset [a, d_a) \cup (c_b, b] \cup \bigcup_{1 \le i \le n} (c_{x_i}, d_{x_i}).$$

We now set

$$Y = \left\{ a, d_a, c_b, b, c_{x_1}, \ldots, c_{x_n}, d_{x_1}, \ldots, d_{x_n} \right\}$$

and write $P : a = y_0 < y_1 < \cdots < y_p = b$ for the partition of $[a, b]$ obtained by arranging the elements of Y in increasing order. If $s, t \in (y_{i-1}, y_i)$, then $\|f(s) - f(t)\|_E < \epsilon$ and it follows that there is a step mapping ϕ such that $\|f - \phi\| < \epsilon$. Hence f is a limit of step mappings and so is regulated.   □

A regulated real-valued function is integrable. From the above proposition, piecewise-continuous and monotone functions are regulated and so integrable. However, there are integrable functions which are not regulated.

Exercise 3.6. Show that the real-valued function f defined on the interval $[-1, 1]$ by

$$f(x) = \begin{cases} \sin \frac{1}{x} & x \neq 0 \\ 0 & x = 0 \end{cases}$$

is integrable, but not regulated.

We will now show how it is possible to define the *integral of a regulated mapping* whose image lies in a general Banach space E. We first consider step mappings. If f is a step mapping defined on an interval $[a, b]$, then there is a partition $P : a =$

$x_0 < x_1 < \cdots < x_n = b$ of $[a, b]$ and elements $c_1, \ldots, c_n \in E$ such that $f(x) = c_i$ on the interval (x_{i-1}, x_i). We define the integral of f by

$$\mathcal{I}(f) = \int_a^b f = \sum_{i=1}^n (x_i - x_{i-1})c_i.$$

The mapping $\mathcal{I} : \mathcal{S}([a, b], E) \longrightarrow E$ is linear and

$$\|\mathcal{I}(f)\|_E \le \sum_{i=1}^n (x_i - x_{i-1})\|c_i\|_E \le (b-a)\|f\|,$$

therefore $\mathcal{I}$ is also continuous. If $a < u < b$, then $f : [a, u] \longrightarrow E$ and $f : [u, b] \longrightarrow E$ are step mappings and

$$\int_a^b f = \int_a^u f + \int_u^b f.$$

For $a \le u \le v \le b$, we set

$$\int_u^u f = 0 \quad \text{and} \quad \int_v^u f = -\int_u^v f.$$

With these conventions, for $u, v, w \in [a, b]$ we have the relation

$$\int_u^v f = \int_u^w f + \int_w^v f,$$

which is called *Chasle's Law*.

Suppose now that f is any regulated mapping. Then f is the limit of a sequence of step mappings (f_n). As $\mathcal{I}$ is linear and continuous, the sequence $(\mathcal{I}(f_n))$ is a Cauchy sequence and so has a limit l, because E is a Banach space. If (g_n) is another sequence of step mappings converging to f, then

$$\|f_n - g_n\| \le \|f_n - f\| + \|f - g_n\|,$$

and so the sequence $(f_n - g_n)$ converges to 0 when n goes to infinity. This implies that the sequence $(\mathcal{I}(g_n))$ also has l for limit, so we can define without ambiguity the integral of f by

$$\mathcal{I}(f) = \int_a^b f = \lim \int_a^b f_n,$$

where (f_n) is any sequence of step mappings converging to f. We also use the notation $\int_a^b f(s)\mathrm{d}s$ for $\int_a^b f$. It is easy to see that the mapping $\mathcal{I}$ extended to $\mathcal{R}([a, b], E)$ is linear. As $\mathcal{I}(f_n)$ converges to $\mathcal{I}(f)$ and $\|f_n\|$ converges to $\|f\|$, we have

$$\|\mathcal{I}(f)\|_E \le (b-a)\|f\|,$$

and so $\mathcal{I}$ is also continuous on $\mathcal{R}([a,b], E)$.

Exercise 3.7. This exercise generalizes what we have just seen. Suppose that E is a normed vector space, F a Banach space, S a vector subspace of E and $f : S \longrightarrow F$ a continuous linear mapping. Show that $\bar{S}$ is a vector subspace of E and that f has a unique extension to a continuous linear mapping $\bar{f} : \bar{S} \longrightarrow F$.

Exercise 3.8. Show that Chasle's Law can be extended to the space of regulated mappings.

Exercise 3.9. Show that if f is a regulated mapping defined on the interval $[a,b]$, then

$$\| \int_a^b f(s)\mathrm{d}s\|_E \le \int_a^b \|f(s)\|_E \mathrm{d}s.$$

As with real-valued functions defined on an interval of $\mathbb{R}$, there is a relation between integrals and derivatives. Let E be a Banach space and $f : [a,b] \longrightarrow E$ regulated. Suppose that $c \in [a,b]$ and for $x \in [a,b]$ let us set

$$F(x) = \int_c^x f.$$

Theorem 3.9. *F is a continuous function. If f is continuous at a point $x \in [a,b]$, then F has a derivative at x and $\dot{F}(x) = f(x)$.*

Proof. We have

$$F(x+h) - F(x) = \int_c^{x+h} f - \int_c^x f = \int_x^{x+h} f$$

and so

$$\|F(x+h) - F(x)\|_E \le |h| \sup_{t \in [x,x+h]} \|f(t)\|_E \le |h| \|f\|_E.$$

This implies that F is continuous.

Suppose now that f is continuous at $x \in [a,b]$. Then

$$\frac{F(x+h) - F(x)}{h} - f(x) = \frac{1}{h} \int_x^{x+h} (f(t) - f(x))\mathrm{d}t.$$

This implies that

$$\left\| \frac{F(x+h) - F(x)}{h} - f(x) \right\|_E \le \sup_{t \in [x,x+h]} \|f(t) - f(x)\|_E$$

and it follows that $\dot{F}(x) = f(x)$. $\qquad\qquad\square$

If the mapping F is such that $\dot{F} = f$, then we say that F is a *primitive* of f. Clearly, if F is a primitive of f and $c \in E$, then so is $F + c$. The previous theorem has the following corollary:

Corollary 3.5. *If $f : [a,b] \longrightarrow E$ is continuous, then f has a primitive.*

We saw above that, if a mapping f has a primitive F, then the addition of a constant to F gives us another primitive. The following result shows that there are no other primitives.

Proposition 3.4. *If F and G are primitives of a mapping $f : [a,b] \longrightarrow E$, then there is a constant $c \in E$ such that $G = F + c$.*

Proof. The derivative of $F - G$ has the value 0 at all points $t \in (a,b)$ and so the differential $(F - G)'$ has the value 0. Using Theorem 3.5 we obtain $F - G = c$, where c is constant. However, F and G are continuous, so $F - G = c$ on $[a,b]$, and hence the result. $\square$

Remark. If $f : [a,b] \longrightarrow E$ is continuous, then from Theorem 3.9 we know that the mapping

$$F : [a,b] \longrightarrow E, x \longmapsto \int_a^x f$$

is a primitive of f. If G is a primitive of f, then from Proposition 3.4 there is a constant c such that $G = F + c$. Therefore

$$G(b) - G(a) = F(b) - F(a) = \int_a^b f.$$

This result generalizes the classical fundamental theorem of calculus. We will refer to it also as the *fundamental theorem of calculus*.

Before closing this section we draw attention to an elementary property of the derivative of a mapping whose image lies in a normed vector space. Let F be a normed vector space, I an interval of $\mathbb{R}$, f a mapping from I into F and α a real-valued function defined on I. Then αf is a mapping from I into F.

Proposition 3.5. *If α and f have derivatives at $t \in I$, then αf also has a derivative at t and*

$$\frac{d}{dt}(\alpha f)(t) = \alpha(t)\dot{f}(t) + \dot{\alpha}(t)f(t).$$

Proof. As

$$\alpha(t+h)f(t+h) - \alpha(t)f(t) = \alpha(t+h)f(t+h) - \alpha(t+h)f(t)$$
$$+ \alpha(t+h)f(t) - \alpha(t)f(t)$$

and the dérivatives $\dot{\alpha}(t)$ and $\dot{f}(t)$ existent, we can write

$$\lim_{h \to 0} \frac{1}{h} \big(\alpha(t+h) f(t+h) - \alpha(t) f(t) \big) = \lim_{h \to 0} \alpha(t+h) \frac{1}{h} \big(f(t+h) - f(t) \big)$$

$$+ \lim_{h \to 0} \frac{1}{h} \big(\alpha(t+h) - \alpha(t) \big) f(t)$$

and the result follows. □

3.4 Differentiation under the Integral Sign

We suppose that O is an open subset of a normed vector space E, F a Banach space, $[a,b]$ a closed bounded interval of $\mathbb{R}$ and $f : [a,b] \times O \longrightarrow F$ continuous.

Lemma 3.2. *For each $x_0 \in O$ and $\epsilon > 0$, there is a $\delta > 0$ such that for all $t \in [a,b]$ we have*

$$\|x - x_0\|_E < \delta \implies \|f(t,x) - f(t,x_0)\|_F < \epsilon.$$

Proof. Let $t \in [a,b]$. As f is continuous at (t,x_0), there is a $\delta_t > 0$ such that

$$\|(s,x) - (t,x_0)\|_{\mathbb{R} \times E} < \delta_t \implies \|f(s,x) - f(t,x_0)\|_F < \frac{\epsilon}{2}.$$

The intervals $(t - \delta_t, t + \delta_t)$ form an open cover of the compact interval $[a,b]$. We may extract a finite open subcover: $[a,b] \subset \cup_{i=1}^{n} (t_i - \delta_{t_i}, t_i + \delta_{t_i})$. We set $\delta = \min \delta_{t_i}$. Suppose now that $\|x - x_0\|_E < \delta$ and let us fix $t \in [a,b]$. There is a t_i such that $|t - t_i| < \delta_{t_i}$. We have

$$\|f(t,x) - f(t,x_0)\|_F \leq \|f(t,x) - f(t_i,x_0)\|_F$$

$$+ \|f(t_i,x_0) - f(t,x_0)\|_F < \frac{\epsilon}{2} + \frac{\epsilon}{2} = \epsilon.$$

This ends the proof. □

We now consider the mapping

$$g : O \longrightarrow F, x \longmapsto \int_a^b f(t,x)\mathrm{d}t$$

Proposition 3.6. *The mapping g is continuous.*

Proof. Let us fix $x \in O$. From the previous section (Exercise 3.9)

$$\left\| \int_a^b f(t,x+h) - f(t,x)\mathrm{d}t \right\|_F \leq (b-a) \sup_{t \in [a,b]} \|f(t,x+h) - f(t,x)\|_F.$$

We now fix $\epsilon > 0$. Applying Lemma 3.2 we obtain $\delta > 0$ such that

$$\|h\|_E < \delta \implies \sup_{t \in [a,b]} \|f(t, x+h) - f(t,x)\|_F \leq \epsilon,$$

which implies that

$$\|g(x+h) - g(x)\|_F \leq (b-a)\epsilon$$

for $\|h\|_E < \delta$. It follows that g is continuous at x.   □

Shortly we will add a hypothesis, which will enable us to strengthen this property. However, before doing so, we need a preliminary result.

Proposition 3.7. *Suppose that E is a normed vector space, F a Banach space, $[a, b]$ a closed bounded interval of $\mathbb{R}$ and A a regulated mapping from $[a, b]$ into $\mathcal{L}(E, F)$. If $h \in E$, then $A(t)h$ is a regulated mapping from $[a, b]$ into F and*

$$\int_a^b A(t)h\,dt = \left(\int_a^b A(t)dt \right) h.$$

Proof. Let S be a step mapping from $[a, b]$ into $\mathcal{L}(E, F)$. We have a partition P : $a = x_0 < x_1 < \cdots < x_n = b$ of $[a, b]$ such that $S(t) = l_i \in \mathcal{L}(E, F)$ for $t \in (t_{i-1}, t_i)$. Clearly $S(t)h$ is a step mapping from $[a, b]$ into F and

$$\left(\int_a^b S(t)dt \right) h = \left(\sum_{i=1}^n (t_i - t_{i-1})l_i \right) h$$

$$= \sum_{i=1}^n (t_i - t_{i-1})l_i h$$

$$= \int_a^b S(t)h\,dt.$$

Now let $A(t)$ be any regulated mapping from $[a, b]$ into $\mathcal{L}(E, F)$ and $(S_i(t))$ a sequence of step mappings converging to $A(t)$. Then $(S_i(t)h)$ is a sequence of step mappings converging to $A(t)h$. Hence

$$\left(\int_a^b A(t)dt \right) h = \left(\lim \int_a^b S_i(t)dt \right) h$$

$$= \lim \int_a^b S_i(t)h\,dt$$

$$= \int_a^b A(t)h\,dt.$$

This ends the proof.   □

Now we suppose that the partial differential $\partial_2 f$ is defined and continuous on the set $[a, b] \times O$. Then we have the

Theorem 3.10. *The mapping*

$$g : O \longrightarrow F, x \longmapsto \int_a^b f(t, x)dt$$

is of class C^1 and

$$g'(x) = \int_a^b \partial_2 f(t, x)dt.$$

Proof. Let us fix $x \in O$ and consider the mapping

$$\Phi_x : E \longrightarrow F, h \longmapsto \int_a^b \partial_2 f(t, x)hdt.$$

Φ_x is clearly linear. Also,

$$\|\partial_2 f(t, x)h\|_F \leq |\partial_2 f(t, x)|_{\mathcal{L}(E,F)}\|h\|_E \leq \sup_{t \in [a,b]} |\partial_2 f(t, x)|_{\mathcal{L}(E,F)}\|h\|_E,$$

which implies that

$$\left\| \int_a^b \partial_2 f(t, x)hdt \right\|_F \leq (b - a) \sup_{t \in [a,b]} |\partial_2 f(t, x)|_{\mathcal{L}(E,F)}\|h\|_E.$$

Therefore Φ_x is also continuous. We will now show that $\Phi_x = g'(x)$. First we have

$$\|g(x + h) - g(x) - \Phi_x(h)\|_F = \left\| \int_a^b f(t, x + h) - f(t, x) - \partial_2 f(t, x)hdt \right\|_F$$

$$\leq (b - a) \sup_{t \in [a,b]} \|f(t, x + h) - f(t, x)$$

$$- \partial_2 f(t, x)h\|_F.$$

From Corollary 3.3, for any t

$$\|f(t, x+h) - f(t, x) - \partial_2 f(t, x)h\|_F \leq \sup_{y \in [x, x+h]} |\partial_2 f(t, y) - \partial_2 f(t, x)|_{\mathcal{L}(E,F)}\|h\|_E.$$

Let us fix $\epsilon > 0$. By hypothesis $\partial_2 f$ is continuous, so from Lemma 3.2 we know that there is a $\delta > 0$ such that, for all $t \in [a, b]$

$$\|h\|_E < \delta \implies |\partial_2 f(t, x + h) - \partial_2 f(t, x)|_{\mathcal{L}(E,F)} < \epsilon.$$

Therefore
$$\|g(x+h) - g(x) - \Phi_x(h)\|_F \leq (b-a)\epsilon\|h\|_E.$$

This proves that Φ_x is the differential of g at x. However, from Proposition 3.7, we have
$$\int_a^b \partial_2 f(t,x)h\mathrm{d}t = \left(\int_a^b \partial_2 f(t,x)\mathrm{d}t\right)h$$

and so we may write
$$g'(x) = \int_a^b \partial_2 f(t,x)\mathrm{d}t.$$

To see that g' is continuous it is sufficient to apply Proposition 3.6. $\square$

Chapter 4
Higher Derivatives and Differentials

Let $O \subset \mathbb{R}^n$ be open and f a real-valued function defined on O. If the function $\frac{\partial f}{\partial x_i}$ is defined on O, then we can consider the existence of its partial derivatives. If $\frac{\partial}{\partial x_j}(\frac{\partial f}{\partial x_i})(a)$ exists, then we write for this derivative $\frac{\partial^2 f}{\partial x_j \partial x_i}(a)$ if $i \neq j$, and $\frac{\partial^2 f}{\partial x_i^2}(a)$ if $i = j$. We also write $\partial^2_{ji} f(a)$, or $\partial^2_{ii} f(a)$, in the case where $i = j$. This derivative is called the (j, i)th second partial derivative of f at a. If the given partial derivative is defined for all $x \in O$, then we obtain a real-valued function on O, also called the (j, i)th second partial derivative. If these functions are defined and continuous for all pairs (j, i), then we say that f is of *class C^2*.

Example. Consider the real-valued function f defined on $\mathbb{R}^2$ by $f(x, y) = \sin xy$. We have

$$\frac{\partial f}{\partial x}(x, y) = y \cos xy \qquad \text{and} \qquad \frac{\partial f}{\partial y}(x, y) = x \cos xy$$

for $(x, y) \in \mathbb{R}^2$. Differentiating the functions $\frac{\partial f}{\partial x}$ and $\frac{\partial f}{\partial y}$, we obtain

$$\frac{\partial^2 f}{\partial x^2}(x, y) = -y^2 \sin xy \qquad \text{and} \qquad \frac{\partial^2 f}{\partial y^2}(x, y) = -x^2 \sin xy$$

and also

$$\frac{\partial^2 f}{\partial y \partial x}(x, y) = \cos xy - xy \sin xy = \frac{\partial^2 f}{\partial x \partial y}(x, y).$$

The functions $\frac{\partial^2 f}{\partial x^2}, \frac{\partial^2 f}{\partial y \partial x}, \frac{\partial^2 f}{\partial x \partial y}$ and $\frac{\partial^2 f}{\partial y^2}$ are defined and continuous on $\mathbb{R}^2$ and so f is of class C^2.

We can now consider the partial derivatives of the second partial derivatives of a real-valued function f and so possibly obtain third partial derivatives. If these are all defined everywhere and continuous, then we say that f is of class C^3. Continuing in the same way we obtain the definition of a function of *class C^k* for any $k \geq 1$.

R. Coleman, *Calculus on Normed Vector Spaces*, Universitext,
DOI 10.1007/978-1-4614-3894-6_4, © Springer Science+Business Media New York 2012

We say that continuous functions are of *class* C^0. If a function is of class C^k for all $k \in \mathbb{N}$, then we say that f is of *class* C^∞, or *smooth*.

Exercise 4.1. Show that a polynomial function defined on $\mathbb{R}^n$ is of class C^∞.

In the example above we saw that the partial derivatives $\frac{\partial^2 f}{\partial y \partial x}$ and $\frac{\partial^2 f}{\partial x \partial y}$ had the same expression. In the next section we will see that this was not an accident.

Before closing this section, notice that we extend these definitions to mappings with image in $\mathbb{R}^m$, with $m > 1$, by considering coordinate functions.

4.1   Schwarz's Theorem

Under certain conditions we may change the order of differentiation without affecting the result. For example, we may differentiate a function $f : \mathbb{R}^2 \longrightarrow \mathbb{R}, (x, y) \longmapsto f(x, y)$ first with respect to x then with respect to y, or vice versa, and obtain the same expression. The next result is known as Schwarz's Theorem.

Theorem 4.1. *Let $O \subset \mathbb{R}^2$ be open and $f : O \longrightarrow \mathbb{R}$ be such that the second partial derivatives $\frac{\partial^2 f}{\partial y \partial x}$ and $\frac{\partial^2 f}{\partial x \partial y}$ are defined on O. If these functions are continuous at $(a, b) \in O$, then*

$$\frac{\partial^2 f}{\partial y \partial x}(a, b) = \frac{\partial^2 f}{\partial x \partial y}(a, b).$$

Proof. As O is open, there is an $\epsilon > 0$ such that closed square $S = [a - \epsilon, a + \epsilon] \times [b - \epsilon, b + \epsilon]$ lies in O. Let (h, k) be an element of the closed square $S' = [-\epsilon, \epsilon] \times [-\epsilon, \epsilon]$ and suppose that $h \neq 0$ and $k \neq 0$. We set

$$\Delta(h, k) = f(a + h, b + k) - f(a, b + k) - f(a + h, b) + f(a, b).$$

If $\phi(s) = f(s, b + k) - f(s, b)$, then $\Delta(h, k) = \phi(a + h) - \phi(a)$. The function ϕ is defined and continuous on $[a, a + h]$ and differentiable on $(a, a + h)$. From Theorem 3.1 there exists $\theta \in (0, 1)$ such that

$$\phi(a + h) - \phi(a) = h\dot{\phi}(a + \theta h),$$

or

$$\Delta(h, k) = h\left(\frac{\partial f}{\partial x}(a + \theta h, b + k) - \frac{\partial f}{\partial x}(a + \theta h, b)\right).$$

We now set $\psi(t) = \frac{\partial f}{\partial x}(a + \theta h, t)$. The function ψ is defined and continuous on $[b, b + k]$ and differentiable on $(b, b + k)$. Using Theorem 3.1 again, we see that there exists $\theta' \in (0, 1)$ such that

$$\psi(b + k) - \psi(b) = k\dot{\psi}(b + \theta' k),$$

or

$$\frac{\partial f}{\partial x}(a+\theta h, b+k) - \frac{\partial f}{\partial x}(a+\theta h, b) = k\frac{\partial^2 f}{\partial y \partial x}(a+\theta h, b+\theta' k).$$

Therefore

$$\Delta(h,k) = hk\frac{\partial^2 f}{\partial y \partial x}(a+\theta h, b+\theta' k).$$

As the function $\frac{\partial^2 f}{\partial y \partial x}$ is continuous at (a,b), we have

$$\lim_{(h,k)\to(0,0)} \frac{\Delta(h,k)}{hk} = \frac{\partial^2 f}{\partial y \partial x}(a,b).$$

We can also write $\Delta(h,k) = \phi_1(b+k) - \phi_1(b)$, where $\phi_1(t) = f(a+h,t) - f(a,t)$. Proceeding as above, we obtain

$$\lim_{(h,k)\to(0,0)} \frac{\Delta(h,k)}{hk} = \frac{\partial^2 f}{\partial x \partial y}(a,b)$$

and hence the result. $\qquad\square$

Corollary 4.1. *Let* $O \subset \mathbb{R}^n$ *be open and* $f : O \longrightarrow \mathbb{R}$ *be such that the second partial derivatives* $\frac{\partial^2 f}{\partial x_j \partial x_i}$ *and* $\frac{\partial^2 f}{\partial x_i \partial x_j}$ *are defined on* O. *If these functions are continuous at* $a \in O$, *then*

$$\frac{\partial^2 f}{\partial x_j \partial x_i}(a) = \frac{\partial^2 f}{\partial x_i \partial x_j}(a).$$

Proof. Let $i : \mathbb{R}^2 \longrightarrow \mathbb{R}^n$ be defined by

$$i(u,v) = a + ue_i + ve_j,$$

where e_i (resp. e_j) is the ith (resp. jth) element of the standard basis of $\mathbb{R}^n$. i is continuous and so the set $U = i^{-1}(O)$ is open in $\mathbb{R}^2$. If we set $g = f \circ i$, then g is defined on U, as are the functions $\frac{\partial^2 g}{\partial v \partial u}$ and $\frac{\partial^2 g}{\partial u \partial v}$, with

$$\frac{\partial^2 g}{\partial v \partial u}(u,v) = \frac{\partial^2 f}{\partial x_j \partial x_i}(a+ue_i+ve_j) \quad \text{and} \quad \frac{\partial^2 g}{\partial u \partial v}(u,v) = \frac{\partial^2 f}{\partial x_i \partial x_j}(a+ue_i+ve_j).$$

As

$$\frac{\partial^2 g}{\partial v \partial u} = \frac{\partial^2 f}{\partial x_j \partial x_i} \circ i \quad \text{and} \quad \frac{\partial^2 g}{\partial u \partial v} = \frac{\partial^2 f}{\partial x_i \partial x_j} \circ i,$$

the functions $\frac{\partial^2 g}{\partial v \partial u}$ and $\frac{\partial^2 g}{\partial u \partial v}$ are continuous at $(0,0)$ and so

$$\frac{\partial^2 g}{\partial v \partial u}(0,0) = \frac{\partial^2 g}{\partial u \partial v}(0,0)$$

and the result follows. $\qquad\qquad\qquad\qquad\qquad\qquad\qquad\qquad\qquad\qquad\square$

Example. Consider the real-valued function f defined on $\mathbb{R}^2$ by

$$f(x,y) = \begin{cases} \frac{xy(x^2-y^2)}{x^2+y^2} & (x,y) \neq (0,0) \\ 0 & (x,y) = (0,0) \end{cases}.$$

It is not difficult to see that f is continuous. Simple calculations show that

$$\frac{\partial f}{\partial x}(0,y) = -y \qquad \text{and} \qquad \frac{\partial f}{\partial y}(x,0) = x$$

and so

$$\frac{\partial^2 f}{\partial y \partial x}(0,0) = -1 \qquad \text{and} \qquad \frac{\partial^2 f}{\partial x \partial y}(0,0) = 1.$$

It is clear that outside of the set $\{(0,0)\}$ the second derivatives exist. Therefore, from Schwarz's Theorem, at least one of the functions $\frac{\partial^2 f}{\partial y \partial x}$ or $\frac{\partial^2 f}{\partial x \partial y}$ is discontinuous at $(0,0)$.

The next result follows directly from what precedes and needs no proof.

Theorem 4.2. *If O is an open subset of $\mathbb{R}^n$, $f : O \longrightarrow \mathbb{R}$ is of class C^2 and $a \in O$, then for $i,j = 1,\ldots,n$ we have*

$$\frac{\partial^2 f}{\partial x_j \partial x_i}(a) = \frac{\partial^2 f}{\partial x_i \partial x_j}(a).$$

It is possible to extend the preceding result to functions of class C^k with $k \geq 3$. Let us write $\frac{\partial^k f}{\partial x_{i_1} \ldots \partial x_{i_k}}$ for the partial derivative of f obtained by differentiating with respect to the variable x_{i_k}, then with respect to the variable $x_{i_{k-1}}$ and so on up to the variable x_{i_1}. We write S_k for the group of permutations of the set $\{1,\ldots,k\}$. Then we have the following result, which we will prove further on:

Theorem 4.3. *Let O be an open subset of $\mathbb{R}^n$, $f : O \longrightarrow \mathbb{R}$ of class C^k and $(i_1,\ldots,i_k) \in \mathbb{N}^k$ with $1 \leq i_1,\ldots,i_k \leq n$. If $\sigma \in S_k$, then for $a \in O$ we have*

$$\frac{\partial^k f}{\partial x_{i_1} \ldots \partial x_{i_k}}(a) = \frac{\partial^k f}{\partial x_{i_{\sigma(1)}} \ldots \partial x_{i_{\sigma(k)}}}(a).$$

4.2   Operations on C^k-Mappings

In this section we will consider the sum, product and so on of mappings of class C^k. We have already handled this subject for the case $k = 1$; here we will extend the results to $k > 1$.

Proposition 4.1. *Let $O \subset \mathbb{R}^n$ be open, $f, g : O \longrightarrow \mathbb{R}$ of class C^k and $\lambda \in \mathbb{R}$. Then the functions $f + g$, λf and fg are of class C^k. If g does not vanish on O, then $h = \frac{f}{g}$ is of class C^k.*

Proof. We have already proved the result for $k = 1$. Suppose that the result is true for a given k and that the functions f and g are of class C^{k+1}. We have

$$\frac{\partial(f + g)}{\partial x_i}(x) = \frac{\partial f}{\partial x_i}(x) + \frac{\partial g}{\partial x_i}(x) \qquad \text{and} \qquad \frac{\partial(\lambda f)}{\partial x_i} = \lambda \frac{\partial f}{\partial x_i}$$

and also

$$\frac{\partial(fg)}{\partial x_i}(x) = g(x)\frac{\partial f}{\partial x_i} + f(x)\frac{\partial g}{\partial x_i}(x).$$

As the functions f, g, $\frac{\partial f}{\partial x_i}$ and $\frac{\partial g}{\partial x_i}$ are of class C^k, so are the functions $\frac{\partial(f+g)}{\partial x_i}$, $\frac{\partial(\lambda f)}{\partial x_i}$ and $\frac{\partial(fg)}{\partial x_i}$. It follows that $f + g$, λf and fg are of class C^{k+1}. If g does not vanish, then

$$\frac{\partial h}{\partial x_i}(x) = \frac{g(x)\frac{\partial f}{\partial x_i}(x) - f(x)\frac{\partial g}{\partial x_i}(x)}{g(x)^2},$$

therefore $\frac{\partial h}{\partial x_i}$ is of class C^k and so h is of class C^{k+1}. The result now follows by induction. $\qquad\square$

Corollary 4.2. *Let $O \subset \mathbb{R}^n$ be open, $f, g : O \longrightarrow \mathbb{R}^m$ of class C^k and $\lambda \in \mathbb{R}$. Then the mappings $f + g$ and λf are of class C^k.*

Proof. As f and g are of class C^k, so are their coordinate functions f_i and g_i, for all i. Now $(f + g)_i = f_i + g_i$ and $(\lambda f)_i = \lambda f_i$, hence $(f + g)_i$ and $(\lambda f)_i$ are of class C^k. It follows that $f + g$ and λf are of class C^k. $\qquad\square$

Remark. We have shown that the C^k-mappings defined on open subset of $\mathbb{R}^n$, with image in $\mathbb{R}^m$, form a vector space. If $m = 1$, then this space is an algebra.

We now consider the composition of mappings.

Proposition 4.2. *Let $O \subset \mathbb{R}^n$, $U \subset \mathbb{R}^m$ be open and $f : O \longrightarrow \mathbb{R}^m$, $g : U \longrightarrow \mathbb{R}^s$ of class C^k with $f(O) \subset U$. Then $g \circ f : O \longrightarrow \mathbb{R}^s$ is of class C^k.*

Proof. We have already proved the result for $k = 1$. Suppose that the result is true for some k and that the functions f and g are of class C^{k+1}. For $x \in O$ we have

$$J_{g \circ f}(x) = J_g(\tilde{f}(x)) \circ J_f(x),$$

therefore

$$\frac{\partial (g \circ f)_i}{\partial x_j}(x) = \sum_{k=1}^{m} \frac{\partial g_i}{\partial y_k}(f(x)) \frac{\partial f_k}{\partial x_j}(x).$$

As $\frac{\partial g_i}{\partial y_k}$ and f are of class C^k, $\frac{\partial g_i}{\partial y_k} \circ f$ is of class C^k. From Proposition 4.1, $\frac{\partial (g \circ f)_i}{\partial x_j}$ is of class C^k and hence $g \circ f$ is of class C^{k+1}. The result follows by induction. $\square$

4.3 Multilinear Mappings

Second and higher differentials are more difficult to define than second and higher derivatives. The natural way of defining a second differential would be to take the differential of the mapping $x \longmapsto f'(x)$. Unfortunately, if E and F are normed vector spaces and f a differentiable mapping from an open subset of E into F, then the image of f' lies not in F but in $\mathcal{L}(E, F)$. This means that the differential of f' lies in $\mathcal{L}(E, \mathcal{L}(E, F))$. If we take third and higher differentials, then the mappings obtained become unwieldy. We get around this problem by identifying differentials with multilinear mappings. In this section we will therefore look into such mappings, before handling higher differentials in the next section.

Let $E_1, \ldots, E_k$ and F be vector spaces and f a mapping from $E_1 \times \cdots \times E_k$ into F. We may fix $k - 1$ coordinates and so obtain a mapping from an E_i into F. If such a mapping is linear for each E_i, then f is said to be k-linear. A 1-linear mapping is just a linear mapping. We use the general term *multilinear mapping* for any k-linear mapping. In the case where $k = 2$, we say that f is bilinear and, in the case where $k = 3$, trilinear. If $E = E_1 = \cdots = E_k$, then we speak of a k-linear mapping from E into F. If $F = \mathbb{R}$, then we use the term *multilinear form* for multilinear mapping.

Let us now suppose that the vector spaces $E_1, \ldots, E_k$ and F are normed vector spaces. Setting $E = E_1 \times \cdots \times E_k$, then as usual we define a norm on E by

$$\|(x_1, \ldots, x_k)\|_E = \max(\|x_1\|_{E_1}, \ldots, \|x_k\|_{E_k}).$$

We can characterize continuous multilinear mappings in a way analogous to that used to characterize linear mappings (Proposition 1.15). We will write $\bar{B}$ for the closed unit ball of E.

Proposition 4.3. *The following statements are equivalent:*

(a) *f is continuous on E;*
(b) *f is continuous at 0;*
(c) *f is bounded on $\bar{B}$;*
(d) *There exists $\mu \in \mathbb{R}_+$ such that*

$$\|f(x_1, \ldots, x_k)\|_F \leq \mu \|x_1\|_{E_1} \cdots \|x_k\|_{E_k}.$$

Proof. (a) $\implies$ (b) This is true by the definition of continuity.

(b) $\implies$ (c) By hypothesis there exists $\alpha > 0$ such that $\|f(x)\|_F \leq 1$ if $\|x\|_E \leq \alpha$. If $y \in \bar{B}$, then $\|\alpha y\|_E \leq \alpha$ and so

$$\alpha^k \|f(y)\|_F = \|\alpha^k f(y)\|_F = \|f(\alpha y)\|_F \leq 1.$$

Hence $\|f(y)\|_F \leq \frac{1}{\alpha^k}$.

(c) $\implies$ (d) By hypothesis there exists $\mu \in \mathbb{R}_+$ such that $\|f(y)\|_F \leq \mu$ if $\|y\|_E \leq 1$. If $x = (x_1, \ldots, x_k) \in E$, with $x_i \neq 0$ for all i, then $y = (\frac{x_1}{\|x_1\|_{E_1}}, \ldots, \frac{x_k}{\|x_k\|_{E_k}}) \in \bar{B}$ and so

$$\|f(y)\|_F \leq \mu \qquad \text{or} \qquad \frac{\|f(x)\|_F}{\|x_1\|_{E_1} \cdots \|x_k\|_{E_k}} \leq \mu.$$

Therefore $\|f(x)\|_F \leq \mu \|x_1\|_{E_1} \cdots \|x_k\|_{E_k}$. Clearly this is also true when one or more of the x_i is equal to 0.

(d) $\implies$ (a) Let us fix $a \in E$ and take $h \in E$. We set

$$h^0 = (0, \ldots, 0) \qquad \text{and} \qquad h^i = (h_1, \ldots, h_i, 0, \ldots 0)$$

for $i = 1, \ldots, k$. Then

$$f(a+h) - f(a) = \sum_{i=1}^{k} f(a+h^i) - f(a+h^{i-1}).$$

However,

$$f(a+h^1) - f(a+h^0) = f(h_1, a_2, \ldots, a_k),$$

for $2 \leq i \leq k-1$,

$$f(a+h^i) - f(a+h^{i-1}) = f(a_1+h_1, \ldots, a_{i-1}+h_{i-1}, h_i, a_{i+1}, \ldots, a_k)$$

and

$$f(a+h^k) - f(a+h^{k-1}) = f(a_1+h_1, \ldots, a_{k-1}+h_{k-1}, h_k).$$

Therefore

$$\|f(a+h) - f(a)\|_F \leq \mu \|h_1\|_{E_1} \|a_2\|_{E_2} \cdots \|a_k\|_{E_k}$$

$$+ \mu \sum_{i=2}^{k-1} \|a_1+h_1\|_{E_1} \cdots \|a_{i-1}+h_{i-1}\|_{E_{i-1}} \|h_i\|_{E_i} \|a_{i+1}\|_{E_{i+1}} \cdots \|a_k\|_{E_k}$$

$$+ \mu \|a_1+h_1\|_{E_1} \cdots \|a_{k-1}+h_{k-1}\|_{E_{k-1}} \|h_k\|_{E_k}.$$

If $\|h\|_E \leq \delta$, then $\|h_i\|_{E_i} \leq \delta$ for all i. This implies that $\|a_i + h_i\|_{E_i} \leq \|a_i\|_{E_i} + \delta$. The continuity of f at a now follows. $\qquad\qquad\qquad\qquad\qquad\qquad$ $\square$

The continuous k-linear mappings from $E = E_1 \times \cdots \times E_k$ into F form a vector space, which we will write $\mathcal{L}(E_1 \times \cdots \times E_k; F)$, or $\mathcal{L}(E; F)$. If $f \in \mathcal{L}(E; F)$ and we set

$$|f| = \sup_{\|x\|_E \leq 1} \|f(x)\|_F,$$

then we obtain a norm. If $E_i \neq \{0\}$ for at least one i, then

$$|f| = \sup_{\|x\|_E = 1} \|f(x)\|_F.$$

Notice that
$$\|f(x_1, \ldots, x_k)\|_F \leq |f|\|x_1\|_{E_1} \cdots \|x_k\|_{E_k}.$$

If $E_1 = \cdots = E_k = E$, then we will write $\mathcal{L}(E^k; F)$ for $\mathcal{L}(E \times \cdots \times E; F)$. The space $\mathcal{L}(E^k; F)$ is not in general the same as the space $\mathcal{L}(E^k, F)$, which is composed of the continuous linear mappings from E^k into F.

Exercise 4.2. Let $E_1, \ldots, E_k$ be finite-dimensional normed vector spaces and $E = E_1 \times \cdots \times E_k$. Show that any multilinear mapping from E into a normed vector space F is continuous.

Exercise 4.3. Let f be a k-linear form on $\mathbb{R}^n$. Show that f may be written

$$f(x_1, \ldots x_k) = \sum a_{i_1, \ldots, i_k} x_{1, i_1} \cdots x_{k, i_k}$$

where the sum is taken over all sequences $(i_1, \ldots, i_k)$, such that $i_j \in \{1, \ldots, n\}$ and the $a_{i_1, \ldots, i_k}$ are real constants. Show that $\sum_{i_1, \ldots, i_k} |a_{i_1, \ldots, i_k}|$ defines a norm on $\mathcal{L}((\mathbb{R}^n)^k; \mathbb{R})$.

Exercise 4.4. Let $E_1, \ldots, E_k$ and $F_1, \ldots, F_p$ be normed vector spaces. We set $E = E_1 \times \cdots \times E_k$ and $F = F_1 \times \cdots \times F_p$. For $f \in \mathcal{L}(E; F)$, we define the mapping

$$\Phi : \mathcal{L}(E; F) \longrightarrow \mathcal{L}(E; F_1) \times \cdots \times \mathcal{L}(E; F_p), f \longmapsto (f_1, \ldots f_p),$$

where $f_1, \ldots, f_p$ are the coordinate mappings of f. Show that Φ is an isometric isomorphism, if E and F have their usual product spaces norms.

Consider two normed vector spaces E and F and let us write $\mathcal{L}_1(E, F)$ for $\mathcal{L}(E, F)$, $\mathcal{L}_2(E, F)$ for $\mathcal{L}(E, \mathcal{L}_1(E, F))$ and so on. We will define linear continuous mappings Φ_k from $\mathcal{L}_k(E, F)$ into $\mathcal{L}(E^k; F)$. We let Φ_1 be the identity on $\mathcal{L}(E, F)$ and, for $k = 2$, we set

$$\Phi_2(f)(x_1, x_2) = f(x_1)x_2 \in F.$$

There is no difficulty in seeing that $\Phi_2(f)$ is 2-linear. Also,

$$\|f(x_1)x_2\|_F \leq |f(x_1)|_{\mathcal{L}_1(E,F)}\|x_2\|_E \leq |f|_{\mathcal{L}_2(E,F)}\|x_1\|_E\|x_2\|_E,$$

hence $\Phi_2(f)$ is continuous. Therefore $\Phi_2(f) \in \mathcal{L}(E^2; F)$. The mapping Φ_2 is linear and continuous with

$$|\Phi_2(f)|_{\mathcal{L}(E^2;F)} \leq |f|_{\mathcal{L}_2(E,F)}.$$

For $k = 3$ we set

$$\Phi_3(f)(x_1, x_2, x_3) = (f(x_1)x_2)x_3 \in F$$

and show, in a similar way, that Φ_3 is a linear continuous mapping from $\mathcal{L}_3(E, F)$ into $\mathcal{L}(E^3; F)$ with

$$|\Phi_3(f)|_{\mathcal{L}(E^3;F)} \leq |f|_{\mathcal{L}_3(E,F)}.$$

For higher values of k we proceed in an analogous way, in each case obtaining

$$|\Phi_k(f)|_{\mathcal{L}(E^k;F)} \leq |f|_{\mathcal{L}_k(E,F)}.$$

Exercise 4.5. Give the details of the construction of $\Phi_3(f)$ for $f \in \mathcal{L}_3(E, F)$.

We may also define a mapping Ψ_k from $\mathcal{L}(E^k; F)$ into $\mathcal{L}_k(E, F)$. This is a little more difficult. As for Φ_1, we let Ψ_1 be the identity on $\mathcal{L}(E, F)$. Suppose now that $g \in \mathcal{L}(E^2; F)$. If we fix $x_1 \in E$, then the mapping $\bar{g}(x_1) : x_2 \longmapsto g(x_1, x_2)$ is linear. In addition,

$$\|\bar{g}(x_1)x_2\|_F \leq |g|_{\mathcal{L}(E^2;F)}\|x_1\|_E\|x_2\|_E.$$

Therefore $\bar{g}(x_1) \in \mathcal{L}(E, F)$ and

$$|\bar{g}(x_1)|_{\mathcal{L}(E,F)} \leq |g|_{\mathcal{L}(E^2;F)}\|x_1\|_E.$$

Now, the mapping

$$\bar{g} : E \longrightarrow \mathcal{L}(E, F), x_1 \longmapsto \bar{g}(x_1)$$

is linear and continuous, with

$$|\bar{g}|_{\mathcal{L}_2(E,F)} \leq |g|_{\mathcal{L}(E^2;F)}.$$

If we set $\Psi_2(g) = \bar{g}$, then Ψ_2 is a linear continuous mapping, with

$$|\Psi_2(g)|_{\mathcal{L}_2(E,F)} \leq |g|_{\mathcal{L}(E^2;F)}.$$

For higher values of k we proceed in an analogous way, in each case obtaining

$$|\Psi_k(g)|_{\mathcal{L}_k(E,F)} \le |g|_{\mathcal{L}(E^k;F)}.$$

Exercise 4.6. Give the details of the construction of $\Psi_3(g)$ for $g \in \mathcal{L}(E^3; F)$.

Theorem 4.4. *If E and F are normed vector spaces, then the mappings Φ_k from $\mathcal{L}_k(E, F)$ into $\mathcal{L}(E^k; F)$ defined above are isometric isomorphisms.*

Proof. It is sufficient to notice that Ψ_k is the inverse of Φ_k.   □

We will refer to the mappings Φ_k and Ψ_k as standard isometric isomorphisms.

Exercise 4.7. Show that, if E and F are normed vector spaces and F complete, then for all $k \ge 1$ $\mathcal{L}(E^k; F)$ is complete.

We say that a k-linear mapping f between vector spaces E and F is *symmetric* if for any permutation $\sigma \in S_k$ we have

$$f(x_1, \ldots, x_k) = f(x_{\sigma(1)}, \ldots, x_{\sigma(k)})$$

for $x_1, \ldots, x_k \in E$. In particular, if f is bilinear and symmetric, then

$$f(x, y) = f(y, x)$$

for all $x, y \in E$. The symmetric k-linear mappings from E into F clearly form a vector subspace of the vector space of k-linear mappings from E into F.

Let us now consider continuous symmetric multilinear mappings. It is easy to see that the continuous symmetric k-linear mappings from E into F form a vector subspace of $\mathcal{L}(E^k; F)$, which we will note $\mathcal{L}_S(E^k; F)$. In fact, we can say a little more.

Proposition 4.4. *The space $\mathcal{L}_S(E^k; F)$ is closed in $\mathcal{L}(E^k; F)$.*

Proof. Let (f_n) be a sequence in $\mathcal{L}_S(E^k; F)$ converging to $f \in \mathcal{L}(E^k; F)$. If $x_1, \ldots, x_k \in E$ and $\sigma \in S_k$, then using the symmetry of f_n we have

$$\begin{aligned}
\|f(x_1, \ldots, x_k) - f(x_{\sigma(1)}, \ldots, x_{\sigma(k)})\| &\le \|f(x_1, \ldots, x_k) - f_n(x_1, \ldots, x_k)\| \\
&+ \|f_n(x_{\sigma(1)}, \ldots, x_{\sigma(k)}) \\
&- f(x_{\sigma(1)} \ldots, x_{\sigma(k)})\| \\
&\le |f - f_n| \|x_1\| \cdots \|x_k\| \\
&+ |f - f_n| \|x_{\sigma(1)}\| \cdots \|x_{\sigma(k)}\| \\
&= 2|f - f_n| \|x_1\| \cdots \|x_k\|.
\end{aligned}$$

It now follows easily that

$$f(x_1, \ldots, x_k) = f(x_{\sigma(1)} \ldots, x_{\sigma(k)}).$$

This ends the proof.   □

Remark. If F is a Banach space, then so is $\mathcal{L}(E^k; F)$ (see Exercise 4.7). As a closed subspace of a Banach space is itself a Banach space, it follows that $\mathcal{L}_S(E^k; F)$ is a Banach space if F is a Banach space.

In the next section we will apply our work here to the study of the differential.

4.4 Higher Differentials

Let E and F be normed vector spaces and O an open subset of E. If $f : O \longrightarrow F$ is differentiable on an open neighbourhood V of $a \in O$, then the mapping

$$f' : V \longmapsto \mathcal{L}(E, F), x \longmapsto f'(x)$$

is defined. As we said in the previous section, if f' is differentiable at a, then we would be tempted to define the second differential $f^{(2)}(a)$ of f at a as $f''(a) = (f')'(a)$. However, in this way $f^{(2)}(a) \in \mathcal{L}_2(E, F)$ and, if we continued the process, we would find $f^{(k)}(a) \in \mathcal{L}_k(E, F)$, spaces which are not easy to handle. Hence we proceed in a different way.

If f' is defined on some neighbourhood of $a \in E$ and its differential $f''(a) \in \mathcal{L}_2(E, F)$ exists, then we say that f is 2-times differentiable (or 2-differentiable) at a and we define the second differential $f^{(2)}(a)$ of f at a to be the 2-linear mapping $\Phi_2(f''(a)) \in \mathcal{L}(E^2; F)$, where Φ_2 is the standard isometric isomorphism from $\mathcal{L}_2(E, F)$ onto $\mathcal{L}(E^2; F)$, which we defined in the previous section. Continuing in the same way we define k-*differentiability* and the kth differential $f^{(k)}(a)$ for higher values of k. We will sometimes write $f^{(1)}$ for f'. To distinguish the differential in $\mathcal{L}_k(E, F)$ corresponding to $f^{(k)}(a)$, we will write $f^{[k]}(a)$ for it, i.e., $\Phi_k(f^{[k]}(a)) = f^{(k)}(a)$.

Proposition 4.5. *Let E and F be normed vector spaces, O an open subset of E and f a mapping from O into F. Then f is $(k + 1)$-differentiable at $a \in O$ if and only if $f^{(k)}$ is differentiable at a and in this case*

$$f^{(k)'}(a)h(h_1, \ldots, h_k) = f^{(k+1)}(a)(h, h_1, \ldots, h_k)$$

for $h, h_1, \ldots, h_k \in E$.

Proof. Suppose first that $f^{(k)}$ is differentiable at a. Then $f^{(k)}$ is defined on a neighbourhood U of a and we have $f^{[k]}(x) = \Psi_k \circ f^{(k)}(x)$ for $x \in U$. It follows that $f^{[k+1]}(a)$ exists.

On the other hand, suppose that f is $(k+1)$-differentiable at a. Then $f^{[k]}$ is defined on a neighbourhood V of a and we have $f^{(k)}(x) = \Phi_k \circ f^{[k]}(x)$ for $x \in V$. This implies that $f^{(k)'}(a)$ exists.

As $f^{(k)'}(a) = \Phi_k \circ f^{[k+1]}(a)$, we have

$$
\begin{aligned}
f^{(k)'}(a)h(h_1, \ldots, h_k) &= \Phi_k(f^{[k+1]}(a)h)(h_1, \ldots, h_k) \\
&= \Phi_{k+1} \circ f^{[k+1]}(a)(h, h_1, \ldots, h_k) \\
&= f^{(k+1)}(h, h_1, \ldots, h_k).
\end{aligned}
$$

This ends the proof.   □

We will now see that a k-linear form $f^{(k)}(a)$ is in fact symmetric. First we will handle the case where $k = 2$.

Theorem 4.5. *Let E and F be normed vector spaces, O an open subset of E and f a mapping from O into F. If f is 2-differentiable at $a \in O$, then $f^{(2)}(a)$ is a symmetric bilinear mapping.*

Proof. For $h, k \in E$ small we set

$$
\Delta(h, k) = f(a + h + k) - f(a + k) - f(a + h) + f(a).
$$

Then

$$
\begin{aligned}
\Delta(h, k) - f^{(2)}(a)(h, k) = {}& f(a + h + k) - f(a + k) - f'(a + h)k + f'(a)k \\
& - (f(a + h) - f(a)) \\
& + (f'(a + h) - f'(a) - f^{[2]}(a)(h))\, k.
\end{aligned}
$$

Now we fix h and set

$$
H(k) = f(a + h + k) - f(a + k) - f'(a + h)k + f'(a)k.
$$

Then

$$
\begin{aligned}
\|\Delta(h, k) - f^{(2)}(a)(h, k)\|_F \leq {}& \|H(k) - H(0)\|_F + |f'(a + h) - f'(a) \\
& - f^{[2]}(a)(h)|_{\mathcal{L}(E,F)} \|k\|_E \\
\leq {}& \sup_{0 \leq \lambda \leq 1} |H'(\lambda k)|_{\mathcal{L}(E,F)} \|k\|_E + |f'(a + h) - f'(a) \\
& - f^{[2]}(a)(h)|_{\mathcal{L}(E,F)} \|k\|_E,
\end{aligned}
$$

from the mean value inequality (Corollary 3.2). However,

$$
\begin{aligned}
H'(u) &= f'(a+h+u) - f'(a+u) - f'(a+h) + f'(a) \\
&= f'(a+h+u) - f'(a) - f^{[2]}(a)(h+u) \\
&\quad - \big(f'(a+u) - f'(a) - f^{[2]}(a)u\big) \\
&\quad - \big(f'(a+h) - f'(a) - f^{[2]}(a)h\big).
\end{aligned}
$$

Now, writing $|\cdot|$ for $|\cdot|_{\mathcal{L}(E,F)}$, we obtain

$$
\begin{aligned}
|H'(u)| &\leq |f'(a+h+u) - f'(a) - f^{[2]}(a)(h+u)| \\
&\quad + |(f'(a+u) - f'(a) - f^{[2]}(a)u)| \\
&\quad + |(f'(a+h) - f'(a) - f^{[2]}(a)h)|.
\end{aligned}
$$

Let us fix $\epsilon > 0$. If h and u are sufficiently small, then we have

$$
|H'(u)| \leq \epsilon \|h+u\|_E + \epsilon \|u\|_E + \epsilon \|h\|_E \leq 2\epsilon(\|h\|_E + \|u\|_E).
$$

If $\lambda \in [0, 1]$, then

$$
|H'(\lambda k)| \leq 2\epsilon(\|h\|_E + \|\lambda k\|_E) \leq 2\epsilon(\|h\|_E + \|k\|_E)
$$

and so

$$
\sup_{0 \leq \lambda \leq 1} |H'(\lambda k)| \leq 2\epsilon(\|h\|_E + \|k\|_E).
$$

Therefore

$$
\begin{aligned}
\|\Delta(h,k) - f^{(2)}(a)(h,k)\|_F &\leq \big(2\epsilon(\|h\|_E + \|k\|_E) + \epsilon\|h\|_E\big)\|k\|_E \\
&= \epsilon(3\|h\|_E + 2\|k\|_E)\|k\|_E \\
&\leq 2\epsilon(\|h\|_E + \|k\|_E)^2.
\end{aligned}
$$

Now,

$$
\begin{aligned}
\|f^{(2)}(a)(h,k) - f^{(2)}(a)(k,h)\|_F &\leq \|f^{(2)}(a)(h,k) - \Delta(h,k)\|_F \\
&\quad + \|\Delta(h,k) - f^{(2)}(a)(k,h)\|_F.
\end{aligned}
$$

As $\Delta(k,h) = \Delta(h,k)$, we obtain

$$
\|f^{(2)}(a)(h,k) - f^{(2)}(a)(k,h)\|_F \leq 4\epsilon(\|h\|_E + \|k\|_E)^2.
$$

Suppose now that $x, y \in E \setminus \{0\}$. If $\alpha > 0$ is small, then so are $h = \frac{\alpha}{2} \frac{x}{\|x\|_E}$ and $k = \frac{\alpha}{2} \frac{y}{\|y\|_E}$ and so

$$\frac{\alpha^2}{4\|x\|_E\|y\|_E} \|f^{(2)}(a)(x, y) - f^{(2)}(a)(y, x)\|_F \leq 4\epsilon\alpha^2,$$

which gives us

$$\|f^{(2)}(a)(x, y) - f^{(2)}(a)(y, x)\|_F \leq 16\epsilon\|x\|_E\|y\|_E,$$

and hence the result.   □

We now turn to the general case.

Corollary 4.3. *Let E and F be normed vector spaces and O an open subset of E. If $f : O \longrightarrow F$ is k-times differentiable at $a \in O$, then $f^{(k)}(a)$ is symmetric.*

Proof. The question only arises for $k \geq 2$ and we have already handled the case $k = 2$. Suppose that the result is true up to a given k and consider the case $k + 1$. If f is $(k+1)$-differentiable at $a \in E$, then $f^{(k)}$ is defined on a neighbourhood V of a and by hypothesis $f^{(k)}(x) \in \mathcal{L}_S(E^k; F)$ for all $x \in V$. Now, $A = \Psi_k(\mathcal{L}_S(E^k; F))$ is closed in $\mathcal{L}_k(E, F)$ and so from Proposition 2.8 the image of $f^{[k]'}(a)$ lies in A, i.e., for any x, $\Phi_k(f^{[k]'}(a)x)$ is symmetric. This means that, if we fix the first variable of $f^{(k+1)}(a)$, then the resulting k-linear mapping is symmetric.

Next we notice that $\mathcal{L}_{k+1}(E, F) = \mathcal{L}_2(E, \mathcal{L}_{k-1}(E, F))$. If we set $g = f^{[k-1]}$, then $g^{[2]}(a) = f^{[k+1]}(a)$ and $g^{[2]}(a) \in \mathcal{L}_2(E, \mathcal{L}_{k-1}(E, F))$. Using Theorem 4.5 we have

$$g^{(2)}(a)(x_1, x_2) = g^{(2)}(a)(x_2, x_1)$$

for $x_1, x_2 \in E$ and it follows that

$$f^{(k+1)}(x_1, x_2, x_3, \dots, x_{k+1}) = f^{(k+1)}(x_2, x_1, x_3, \dots, x_{k+1})$$

for $x_1, x_2, x_3, \dots, x_{k+1} \in E$.

Suppose now that $\sigma \in S_{k+1}$ and consider the expression

$$f^{(k+1)}(a)(x_{\sigma(1)}, \dots, x_{\sigma(k+1)}).$$

If $\sigma(1) = 1$, then

$$f^{(k+1)}(a)(x_{\sigma(1)}, x_{\sigma(2)} \dots, x_{\sigma(k+1)}) = f^{(k+1)}(a)(x_1, x_{\sigma(2)} \dots, x_{\sigma(k+1)})$$
$$= f^{(k+1)}(a)(x_1, x_2, \dots, x_{k+1}).$$

Now suppose that $\sigma(1) \neq 1$. By hypothesis we can commute the last k variables to obtain

$$f^{(k+1)}(a)(x_{\sigma(1)}, \dots, x_{\sigma(k+1)}) = f^{(k+1)}(a)(x_{\sigma(1)}, x_1, \dots).$$

However, we can commute the first two variables. This followed by a second commutation of the last k variables gives us

$$f^{(k+1)}(a)(x_{\sigma(1)}, x_1, \ldots) = f^{(k+1)}(a)(x_1, x_{\sigma(1)}, \ldots)$$
$$= f^{(k+1)}(a)(x_1, x_2, \ldots, x_{k+1}).$$

The result follows by induction. $\square$

4.5 Higher Differentials and Higher Derivatives

We will now use the work of the previous section to obtain results on higher derivatives of functions defined on open subsets of $\mathbb{R}^n$.

We recall the definition of the directional derivative. Let O be an open subset of a normed vector space E, f a mapping defined on O whose image lies in a normed vector space F, $a \in O$ and u an element of E. If the limit

$$\lim_{t \longrightarrow 0} \frac{f(a + tu) - f(a)}{t}$$

exists, then this derivative, written $\partial_u f(a)$, is called the directional derivative of f at a in the direction u. If a mapping f is differentiable at a point a, then all its directional derivatives exist and $f'(a)u = \partial_u f(a)$. If F is a product of m normed vector spaces, in particular $\mathbb{R}^m$, then

$$\partial_u f(a) = (\partial_u f_1(a), \ldots, \partial_u f_m(a)),$$

where the f_i are the coordinate mappings of f.

Suppose now that f is 2-differentiable at a. Then, for $v \in E$ we have

$$f^{[2]}(a)v = \frac{\mathrm{d}}{\mathrm{d}t} f'(a + tv)_{|t=0} \in \mathcal{L}(E, F).$$

However,

$$\left(\frac{\mathrm{d}}{\mathrm{d}t} f'(a + tv)_{|t=0} \right) u = \frac{\mathrm{d}}{\mathrm{d}t} (f'(a + tv)u)_{|t=0}.$$

As

$$f'(a + tv)u = \frac{\mathrm{d}}{\mathrm{d}s} f(a + tv + su)_{|s=0},$$

we obtain

$$f^{(2)}(a)(v, u) = \frac{\mathrm{d}}{\mathrm{d}t} \frac{\mathrm{d}}{\mathrm{d}s} f(a + tv + su)_{|t=s=0}.$$

More generally, if f is k-times differentiable and $v_1, \ldots, v_k \in E$, then we have

$$f^{(k)}(a)(v_1, \ldots, v_k) = \frac{d}{dt_1} \cdots \frac{d}{dt_k} f(a + t_1 v_1 + \cdots + t_k v_k)|_{t_1 = \cdots = t_k = 0}.$$

As $f^{(k)}$ is symmetric, changing the order of derivation on the right-hand side does not change the result.

If we now take $E = \mathbb{R}^n$, $F = \mathbb{R}$ and $v_j = e_{i_j}$, for $j = 1, \ldots, k$, where the e_{i_j} are elements of the standard basis of $\mathbb{R}^n$, then we see that

$$\frac{\partial^k f}{\partial x_{i_1} \cdots \partial x_{i_k}}(a) = f^{(k)}(a)(e_{i_1}, \ldots, e_{i_k}),$$

i.e., all kth order partial derivatives exist and permutation of the order of differentiation does not affect the value of the partial derivative. Also, if $h_1, \ldots, h_k \in \mathbb{R}^n$ and $h_i = \sum_{j=1}^n h_{ij} e_j$, then

$$f^{(k)}(a)(h_1, \ldots, h_k) = f^{(k)}(a)\left(\sum_{j=1}^n h_{1j} e_j, \ldots, \sum_{j=1}^n h_{kj} e_j \right)$$

$$= \sum_{i_1, \ldots, i_k} \frac{\partial^k f}{\partial x_{i_1} \cdots \partial x_{i_k}}(a) h_{1, i_1} \cdots h_{k, i_k},$$

where the sum is taken over all sequences $(i_1, \ldots, i_k)$ such that $i_j \in \{1, \ldots, n\}$.

An important case of the above is when $k = 2$. Then the result can be written in matrix form:
$$f^{(2)}(a)(h_1, h_2) = h_1^t \mathcal{H} f(a) h_2,$$
where

$$\mathcal{H} f(a) = \left(\frac{\partial^2 f}{\partial x_i \partial x_j}(a) \right)_{1 \leq i, j \leq n}$$

and we have identified the vectors h_1 and h_2 with column vectors. The symmetric matrix $\mathcal{H} f(a)$ is called the *Hessian matrix* of f at a.

Example. In the example after the corollary to Schwarz's Theorem, the second partial derivatives $\frac{\partial^2 f}{\partial x \partial y}(0, 0)$ and $\frac{\partial^2 f}{\partial y \partial x}(0, 0)$ are not equal; therefore the function cannot be 2-differentiable at the origin.

Let us now consider mappings f from an open interval $I \subset \mathbb{R}$ into a normed vector space F, i.e., curves in F. We have already seen that a curve f is differentiable at a point of $a \in I$ if and only if f has a derivative $\dot{f}(a)$ at a. Also, in this case
$$f'(a)s = s \dot{f}(a).$$

We will now show that the situation is similar for higher derivatives.

Theorem 4.6. *Let $I \in \mathbb{R}$ be an open interval, F a normed vector space, f a mapping from I into $\mathbb{R}$ and $k \in \mathbb{N}^*$. Then f is k-differentiable at a if and only if f has a kth derivative at a. In this case we have*

$$|f^{(k)}(a)| = \left\| \frac{\mathrm{d}^k}{\mathrm{d}t^k} f(a) \right\|.$$

Proof. For $k = 1$ we have already proved the result, so let us consider higher values of k. First suppose that f is k-differentiable at a. We have seen above that, if f is a mapping from an open subset O of a normed vector space E into a normed vector space F and k-differentiable at a point a, then

$$f^{(k)}(a)(v_1, \ldots, v_k) = \frac{\mathrm{d}}{\mathrm{d}t_1} \cdots \frac{\mathrm{d}}{\mathrm{d}t_k} f(a + t_1 v_1 + \cdots + t_k v_k)|_{t_1 = \cdots = t_k = 0}.$$

If we now take $E = \mathbb{R}$, O an open interval of $\mathbb{R}$ and $v_i = 1$, then we obtain

$$f^{(k)}(a)(1, \ldots, 1) = \frac{\mathrm{d}^k}{\mathrm{d}t^k} f(a),$$

and so the kth derivative is defined at a.

Now suppose that the derivative $\frac{\mathrm{d}^k}{\mathrm{d}t^k} f(a)$ exists. We set

$$g(a)(h_1, \ldots, h_k) = h_k \cdots h_1 \frac{\mathrm{d}^k}{\mathrm{d}t^k} f(a),$$

with $h_1, \ldots, h_k \in \mathbb{R}$. We claim that $g(a) = f^{(k)}(a)$. First we have

$$f'(a + h_1)h_2 - f'(a)h_2 = h_2 \dot{f}(a + h_1) - h_2 \dot{f}(a)$$

$$= h_2 \left(\frac{\mathrm{d}^2}{\mathrm{d}t^2} f(a)h_1 + o(h_1) \right).$$

The mapping $h_2 \longmapsto h_2 h_1 \frac{\mathrm{d}^2}{\mathrm{d}t^2} f(a)$ belongs to $\mathcal{L}(\mathbb{R}, F)$. Thus the mapping $h_1 \longmapsto h_1 \frac{\mathrm{d}^2}{\mathrm{d}t^2} f(a)$ is a mapping from $\mathbb{R}$ into $\mathcal{L}(\mathbb{R}, F)$. As this mapping is linear and continuous, we have

$$f^{(2)}(a)(h_1, h_2) = f^{[2]}(a)(h_1)h_2 = h_2 h_1 \frac{\mathrm{d}^2}{\mathrm{d}t^2} f(a).$$

Continuing in the same way we obtain

$$f^{(k)}(a)(h_1, \ldots, h_k) = h_k \cdots h_1 \frac{\mathrm{d}^k}{\mathrm{d}t^k} f(a).$$

This proves that $g(a) = f^{(k)}(a)$, i.e., f is k-differentiable at a. Proving that

$$|f^{(k)}(a)| = \left\| \frac{d^k}{dt^k} f(a) \right\|$$

is elementary.   □

Remark. We may extend the term "k-differentiable" to a mapping defined on any interval I. If $f : I \longrightarrow E$ has derivatives $\frac{d^s}{dt^s} f(a)$ at every point $a \in I$ for $s = 1, \ldots, k$, then we say that f is k-differentiable.

4.6   Cartesian Product Image Spaces

In Chap. 2 we saw that if E and F are normed vector spaces, with F a cartesian product of p normed vector spaces, and f a mapping from an open subset of E into F, then f is differentiable at $a \in O$ if and only if the coordinate mappings $f_1, \ldots, f_p$ are differentiable at a. In this case we can write

$$f'(a) = (f_1'(a), \ldots, f_p'(a)).$$

We will now show that we have an analogous result for higher differentials.

Proposition 4.6. *Let O be an open subset of a normed vector space E and f a mapping from O into a normed vector space F, which is a cartesian product of p normed vector spaces. Then f is k-differentiable at $a \in O$ if and only if the coordinate mappings $f_1, \ldots, f_p$ are k-differentiable at a. In this case we have*

$$f^{(k)}(a) = (f_1^{(k)}(a), \ldots, f_p^{(k)}(a)).$$

Proof. We have already proved the result for $k = 1$, so let us assume that $k \geq 2$.

Suppose that the differentials $f_1^{(k)}(a), \ldots, f_p^{(k)}(a)$ exist. From Proposition 4.5 $f_i^{(k-1)'}(a)$ exists for all i and so $f^{(k-1)'}(a)$ exists. Using Proposition 4.5 again, we see that $f^{(k)}(a)$ exists.

Now suppose that the differential $f^{(k)}(a)$ exists. This implies that $f^{(k-1)'}(a)$ exists and it follows that $f_i^{(k-1)'}(a)$ exists for all i. From this we deduce that $f_i^{(k)}(a)$ exists for all i.

As

$$f^{(k-1)'}(a) = (f_1^{(k-1)'}(a), \ldots, f_p^{(k-1)'}(a)),$$

from Proposition 4.5 we obtain

$$f^{(k)}(a) = (f_1^{(k)}(a), \ldots, f_p^{(k)}(a)).$$

This ends the proof.   □

Remark. The differential $f^{(k)}(a)$ is a mapping from E^k into F. We have shown that the coordinate mappings of $f^{(k)}(a)$ are $f_1^{(k)}(a), \ldots, f_p^{(k)}(a)$, i.e., the coordinate mappings of the kth differential at a are the kth differentials at a of the coordinate mappings of f. Thus f is k-differentiable on an open subset U of E if and only if its coordinate mappings are k-differentiable on U, and in this case we have

$$f^{(k)}(x) = (f_1^{(k)}(x), \ldots, f_p^{(k)}(x))$$

for all $x \in U$. This implies that $f^{(k)}$ is continuous on U if and only if the mappings $f_1^{(k)}, \ldots, f_p^{(k)}$ are continuous on U.

4.7 Higher Partial Differentials

We have seen that a C^1-mapping defined on a cartesian product space has continuous partial differentials (Proposition 3.2). It is natural to ask what can be said in the case where the mapping has a higher class of differentiability. In this section we will study this question.

Let $E_1, \ldots, E_p$ and F be normed vector spaces and let us set $E = E_1 \times \cdots \times E_p$. For $f \in \mathcal{L}(E, F)$ and $i = 1, \ldots, p$, we define $f_i : E_i \longrightarrow F$ by

$$f_i(x) = f(0, \ldots, 0, x, 0 \ldots, 0),$$

where the x is in the ith position. For each i, $f_i \in \mathcal{L}(E_i, F)$ and the mapping $G : f \longmapsto (f_1, \ldots, f_p)$ is a linear mapping from $\mathcal{L}(E, F)$ into $L = \mathcal{L}(E_1, F) \times \cdots \times \mathcal{L}(E_p, F)$. In fact, G is a linear isomorphism with G^{-1} defined by

$$G^{-1}(f_1, \ldots, f_p)(x_1, \ldots, x_p) = f_1(x_1) + \cdots + f_p(x_p).$$

We can say a little more.

Proposition 4.7. *G is a normed vector space isomorphism.*

Proof. We have

$$
\begin{aligned}
|f| &= \sup_{\|(x_1, \ldots, x_p)\| \leq 1} \|f(x_1, \ldots, x_p)\| \\
&= \sup_{\max \|x_i\| \leq 1} \|f(x_1, \ldots, x_p)\| \\
&\leq \sup_{\|x_1\| \leq 1} \|f_1(x_1)\| + \cdots + \sup_{\|x_p\| \leq 1} \|f_p(x_p)\|.
\end{aligned}
$$

However,

$$\|(f_1, \ldots, f_p)\|_L = \max\ (|f_1|, \ldots, |f_p|)$$
$$= \max\ (\sup_{\|x_1\| \leq 1} \|f_1(x_1)\|, \ldots, \sup_{\|x_p\| \leq 1} \|f_p(x_p)\|).$$

Therefore

$$|f| \leq p\|(f_1, \ldots, f_p)\|_L$$

and so G^{-1} is continuous. In addition,

$$\|f_1(x_1)\| = \|f(x_1, 0, \ldots, 0)\| \leq |f|\|(x_1, 0, \ldots, 0)\| = |f|\|x_1\|,$$

hence $|f_1| \leq |f|$. In the same way, for any other i, $|f_i| \leq |f|$ and so

$$\|(f_1, \ldots, f_p)\|_L \leq |f|.$$

Thus G is continuous. We have shown that G is a normed vector space isomorphism. $\square$

We now consider partial differentials.

Theorem 4.7. *If $E_1, \ldots, E_p$ and F are normed vector spaces, O an open subset of $E = E_1 \times \cdots \times E_p$ and f a mapping of class C^k, with $k \geq 1$, from O into F, then the partial differentials $\partial_1 f, \ldots, \partial_p f$ are of class C^{k-1}.*

Proof. For $k = 1$ we have already proved the result (Proposition 3.2), so suppose that $k \geq 2$. If $a \in O$, then $f'(a) \in \mathcal{L}(E, F)$ and

$$f'(a)(x_1, \ldots, x_p) = \partial_1 f(a)x_1 + \cdots + \partial_p f(a)x_p.$$

As

$$f'(a)(0, \ldots, 0, x_i, 0, \ldots, 0) = \partial_i f(a)x_i,$$

we have, in the notation of the previous proposition,

$$G(f'(a)) = (\partial_1 f(a), \ldots, \partial_p f(a)).$$

The mapping $a \longmapsto f'(a)$ is of class C^{k-1} and G continuous (and hence smooth), therefore the mapping $a \longmapsto G(f'(a))$ is of class C^{k-1} and it follows from Proposition 4.6 that the partial differentials $\partial_i f$ are of class C^{k-1}. $\square$

4.8 Generalizing C^k to Normed Vector Spaces

We have seen that a mapping f defined on open subset of $\mathbb{R}^n$ with image in $\mathbb{R}^m$ is of class C^1 if and only if it is differentiable on O and the differential f' is a continuous mapping from O into $\mathcal{L}(\mathbb{R}^n, \mathbb{R}^m)$. Our aim in this section is to show that this result may be extended to mappings of class C^k for any k. In the following we will suppose that $\mathbb{R}^n$ is endowed with the norm $\|\cdot\|_\infty$, which we will write $\|\cdot\|$. We will first consider the case of real-valued functions, i.e., where $m = 1$.

Proposition 4.8. *Let O be an open subset of $\mathbb{R}^n$ and $f : O \longrightarrow \mathbb{R}$ k-differentiable. If $f^{(k)}$ is continuous, then f is of class C^k.*

Proof. As f is k-differentiable, f has all partial derivatives of order k at any point $a \in O$. Given that $f^{(k)}$ is continuous at a, for x sufficiently small and $h_1, \ldots, h_k \in \mathbb{R}^n$ we can write

$$|f^{(k)}(a + x)(h_1, \ldots, h_k) - f^{(k)}(a)(h_1, \ldots, h_k)|$$
$$\leq |f^{(k)}(a + x) - f^{(k)}(a)|_{\mathcal{L}((\mathbb{R}^n)^k; \mathbb{R})} \|h_1\| \cdots \|h_k\|.$$

If we take $h_1 = e_{i_1}, \ldots, h_k = e_{i_k}$, where the e_{i_j} are members of the standard basis of $\mathbb{R}^n$, then we obtain

$$\left| \frac{\partial^k f}{\partial x_{i_1} \cdots \partial x_{i_k}}(a + x) - \frac{\partial^k f}{\partial x_{i_1} \cdots \partial x_{i_k}}(a) \right| \leq |f^{(k)}(a + x) - f^{(k)}(a)|_{\mathcal{L}((\mathbb{R}^n)^k; \mathbb{R})}$$

and so the partial derivatives are continuous at a. We have proved that f is of class C^k.   □

We now turn to the converse of this result. Let f be a real-valued function of class C^k defined on an open subset O of $\mathbb{R}^n$. For $x \in O$ we set

$$f_{(k)}(x)(h_1, \ldots, h_k) = \sum_{i_1, \ldots, i_k} \frac{\partial^k f}{\partial x_{i_1} \cdots \partial x_{i_k}}(x) h_{1,i_1} \ldots h_{k,i_k},$$

where the sum is taken over all distinct sequences $(i_1, \ldots, i_k)$ such that $i_j \in \{1, \ldots, k\}$. $f_{(k)}$ is clearly a k-linear form. We have already seen that $f^{(k)}(a) = f_{(k)}(a)$ if f is k-differentiable at a.

Proposition 4.9. *Let f be a real-valued function defined on an open subset O of $\mathbb{R}^n$. If f is of class C^k, then f is k-differentiable on O, $f^{(k)}(x) = f_{(k)}(x)$ for all $x \in O$, and $f^{(k)}$ is continuous.*

Proof. We will establish this result by induction on k. We have already proved the statement for $k = 1$. Suppose now that it is true for k and consider the case $k + 1$. $f^{(k)}$ is a mapping from O into $\mathcal{L}((\mathbb{R}^n)^k; \mathbb{R})$, which has partial derivatives at all points $x \in O$:

$$\frac{\partial}{\partial x_i} f^{(k)}(x)(h_1, \ldots, h_k) = \sum_{i_1, \ldots, i_k} \frac{\partial^{k+1} f}{\partial x_i \partial x_{i_1} \cdots \partial x_{i_k}}(x) h_{1,i_1} \cdots h_{k,i_k}.$$

As f is of class C^{k+1}, the partial derivatives of $f^{(k)}$ are continuous on O. It follows that $f^{(k)}$ has continuous partial differentials on O:

$$\partial_i f^{(k)}(x)s_i(h_1,\ldots,h_k) = s_i \sum_{i_1,\ldots,i_k} \frac{\partial^{k+1} f}{\partial x_i \partial x_{i_1} \cdots \partial x_{i_k}}(x)h_{1,i_1} \cdots h_{k,i_k}.$$

Hence $f^{(k)}$ is differentiable on O. However, from Proposition 4.5, f is $(k+1)$-differentiable on O and

$$f^{(k)'}(x)s(h_1,\ldots,h_k) = f^{(k+1)}(x)(s,h_1,\ldots,h_k).$$

Thus $f^{(k+1)}$ has the required form and the differential $f^{(k+1)}$ is continuous on O.   □

Example. Multivariate polynomials are of class C^k for every $k \geq 1$ and so are k-differentiable for every $k \geq 1$.

We have shown that, if f is a real-valued function defined on an open subset of $\mathbb{R}^n$, then f is of class C^k if and only if f is k-differentiable and the kth differential is continuous. We now suppose that the image of f lies in $\mathbb{R}^m$, with m not necessarily restricted to the value 1. However, f is k-differentiable at a point x and the kth differential is continuous at x if and only if this is the case for the coordinate mappings. We thus obtain the following more general result:

Theorem 4.8. *If O is an open subset of $\mathbb{R}^n$ and f is a mapping from O into $\mathbb{R}^m$, then f is of class C^k if and only if f is k-differentiable and the mapping $f^{(k)}$ is continuous.*

This theorem suggests the following generalization of the notion of class C^k. If E and F are normed vector spaces, O an open subset of E and f a mapping from O into F, then we say that f is of *class C^k* if f is k-differentiable and the mapping $f^{(k)}$ is continuous. If f is of class C^k for all $k \geq 1$, then we say that f is of *class C^∞ or smooth*. The mappings of class C^k from O into F form a subspace of the vector space of mappings from O into F.

Examples. 1. Let E and F be normed vector spaces and $f : E \longrightarrow F$ a continuous linear mapping. Then $f'(x) = f$ for all $x \in E$. As the mapping $x \longmapsto f'(x) \in \mathcal{L}(E,F)$ is constant, f is of class C^∞ and $f^{(k)} = 0$, for $k \geq 2$.

2. Suppose now that E_1, E_2 and F are normed vector spaces and that $f : E_1 \times E_2 \longrightarrow F$ is a continuous bilinear mapping. f is differentiable and at any point $(x,y) \in E_1 \times E_2$ and we have

$$f'(x,y)(h,k) = f(x,k) + f(h,y)$$

for $(h, k) \in E_1 \times E_2$. The mapping $f' : E_1 \times E_2 \longrightarrow \mathcal{L}(E_1 \times E_2, F)$ is clearly linear and continuous and so $f^{(2)}$ exists and is constant. It follows that f is of class C^∞ and $f^{(k)} = 0$ for $k \geq 3$. If E is a normed vector space, then the addition and scalar multiplication are both bilinear and continuous, therefore of class C^∞.

Exercise 4.8. Let $E_1, \ldots, E_p$ and F be normed vector spaces and $f : E_1 \times \cdots \times E_p \longrightarrow F$ a continuous p-linear mapping. Show that f is of class C^∞ and that $f^{(k)} = 0$ for $k \geq p + 1$.

We now state a result which follows directly from our previous work and the definition of a mapping of class C^k which we have just given.

Proposition 4.10. *Let E and F be normed vector spaces, where F is the Cartesian product of the normed vector spaces $F_1, \ldots, F_p$. If O is an open subset of E and f is a mapping from O into F, then f is of class C^k if and only if the coordinate mappings $f_1, \ldots, f_p$ are all of class C^k.*

Remark. We have seen that we may extend the term "k-differentiable" to a mapping defined on any interval I. We may also extend the term class "C^k". If $f : I \longrightarrow E$ is k-differentiable and the mapping $\frac{d^s}{dt^s} f$ is continuous, then we say that f is of class C^k.

4.9 Leibniz's Rule

If f and g are real-valued functions defined and differentiable on an open interval I of $\mathbb{R}$, then the product fg is differentiable and

$$\frac{d}{dx}(fg)(x) = \frac{df}{dx}(x)g(x) + f(x)\frac{dg}{dx}(x)$$

for $x \in I$. This is known as Leibniz's rule. Clearly, if f and g are of class C^1, then so is fg. In Proposition 4.1 we generalized this: if f and g are of class C^k then so is fg. The function $(u, v) \longmapsto uv$ from $\mathbb{R}^2$ into $\mathbb{R}$ is a continuous and bilinear and so we are naturally led to consider mappings of the form $x \longmapsto b(f(x), g(x))$, where b is a continuous bilinear mapping and f and g mappings of class C^k. The following generalization of Leibniz's rule proves to be particularly useful.

Theorem 4.9. *Let E, F_1, F_2 and G be normed vector spaces, O an open subset of E and $f : O \longrightarrow F_1$, $g : O \longrightarrow F_2$ mappings of class C^k. Suppose also that the mapping $b : F_1 \times F_2 \longrightarrow G$ is a continuous bilinear mapping. Then the mapping $\phi : O \longrightarrow G$ defined by*

$$\phi(x) = b(f(x), g(x))$$

is of class C^k.

Proof. We will prove this result by induction on k. Let us first consider the case $k = 1$. ϕ is a composition of the differentiable mappings $x \longmapsto (f(x), g(x))$ and b, hence ϕ is differentiable and

$$\phi'(x)h = b(f(x), g'(x)h) + b(f'(x)h, g(x)).$$

If we set

$$l_1(x)h = b(f(x), g'(x)h) \quad \text{and} \quad l_2(x)h = b(f'(x)h, g(x)),$$

then $l_1(x), l_2(x) \in \mathcal{L}(E, G)$ and $\phi'(x) = l_1(x) + l_2(x)$. For u sufficiently small we have

$$
\begin{aligned}
l_1(x + u)h - l_1(x)h &= b(f(x + u), g'(x + u)h) - b(f(x), g'(x)h) \\
&= b(f(x) + f'(x)u + o(u), g'(x + u)h) - b(f(x), g'(x)h) \\
&= b(f(x), (g'(x + u) - g'(x))h) + b(f'(x)u, g'(x + u)h) \\
&\quad + b(o(u), g'(x + u)h).
\end{aligned}
$$

If $\|h\| \leq 1$, then

$$
\begin{aligned}
\|l_1(x + u)h - l_1(x)h\| &\leq |b|\|f(x)\|\|g'(x + u) - g'(u)| \\
&\quad + |b|\|f'(x)\|\|u\|\|g'(x + u)| + |b|\|o(u)\|\|g'(x + u)|.
\end{aligned}
$$

As the right-hand side of this expression converges to 0 when u converges to 0, l_1 is continuous at x. In the same way l_2 is continuous at x and so ϕ' is continuous at x. It follows that ϕ is of class C^1.

Suppose now that the result is true up to order k and that f and g are of class C^{k+1}. For $A \in \mathcal{L}(E, F_1)$ and $y \in F_2$, let $l_{A,y} : E \longrightarrow G$ be defined by

$$l_{A,y}(h) = b(A(h), y).$$

Clearly $l_{A,y} \in \mathcal{L}(E, G)$. If we set $\alpha(A, y) = l_{A,y}$, then α is a continuous bilinear mapping from $\mathcal{L}(E, F_1) \times F_2$ into $\mathcal{L}(E, G)$. However, f' and g are of class C^k, therefore by hypothesis the mapping $x \longmapsto \alpha(f'(x), g(x))$ is of class C^k. If we now set, for $y \in F_1$ and $B \in \mathcal{L}(E, F_2)$,

$$l_{y,B}(h) = b(y, B(h)),$$

then $l_{y,B} \in \mathcal{L}(E, G)$. The mapping β defined by $\beta(y, B) = l_{y,B}$ is a continuous bilinear mapping from $F_1 \times \mathcal{L}(E, F_2)$ into $\mathcal{L}(E, G)$. Using the induction hypothesis again, we see that the mapping $x \longmapsto \beta(f(x), g'(x))$ is of class C^k. However,

$$\phi'(x) = \alpha(f'(x), g(x)) + \beta(f(x), g'(x))$$

and so ϕ' is of class C^k. It follows that ϕ is of class C^{k+1}. This finishes the induction step and so the proof. $\qquad\square$

As a first application of this result we will prove that a composition of class C^k-mappings is of class C^k.

Theorem 4.10. *Let E, F and G be normed vector spaces, O an open subset of E, U an open subset of F and $f : O \longrightarrow F$, $g : U \longrightarrow G$ mappings of class C^k with $f(O) \subset U$. Then the mapping $g \circ f : O \longrightarrow G$ is of class C^k.*

Proof. For $A \in \mathcal{L}(E, F)$ and $B \in \mathcal{L}(F, G)$, let $\alpha(A, B) = B \circ A \in \mathcal{L}(E, G)$. Then α is bilinear and continuous. As f and g are differentiable, $g \circ f$ is differentiable and

$$(g \circ f)'(x) = g'(f(x)) \circ f'(x) = \alpha(f'(x), g' \circ f(x)).$$

If $k = 1$, then we see immediately that $(g \circ f)'$ is continuous and hence that $g \circ f$ is of class C^1. If $k \geq 2$, then from the previous theorem $(g \circ f)'$ is of class C^{k-1} and it follows that $g \circ f$ is of class C^k. $\qquad\square$

As a second application of Theorem 4.9, let us return to the inversion mapping $\phi : x \longmapsto x^{-1}$ defined on the open group $E^\times$ of invertible elements of a Banach algebra E. We have already seen that this mapping is a homeomorphism from $E^\times$ onto itself (Theorem 1.10). We now consider the differentiability of ϕ.

Theorem 4.11. *Let E be a Banach algebra and $\phi : E^\times \longrightarrow E$ the inversion mapping, i.e., $\phi(x) = x^{-1}$. Then ϕ is of class C^∞.*

Proof. First let us show that ϕ is differentiable. For $x \in E^\times$ we define the mapping l_x from E into itself by $l_x(h) = -x^{-1}hx^{-1}$. l_x is clearly linear. Also, $\| - x^{-1}hx^{-1} \| \leq \| - x^{-1} \|^2 \|h\|$, and so l_x is continuous. In addition, for h small we have

$$\| \phi(x + h) - \phi(x) - l_x(h) \| = \| (x + h)^{-1} - x^{-1} + x^{-1}hx^{-1} \|$$

$$= \| (x + h)^{-1}(x - (x + h) + (x + h)x^{-1}h)x^{-1} \|$$

$$= \| (x + h)^{-1}(hx^{-1})^2 \|$$

$$\leq \| (x + h)^{-1} \| \|x^{-1}\|^2 \|h\|^2.$$

This shows that ϕ is differentiable, with $\phi'(x) = l_x$.

We will now show that ϕ is of class C^1. For $a, b \in E$ we define an element $\alpha(a, b)$ of $\mathcal{L}(E)$ by setting $\alpha(a, b)(h) = -ahb$ for $h \in E$. α is clearly a bilinear mapping from E^2 into $\mathcal{L}(E)$. As $\|\alpha(a, b)(h)\| \leq \|a\|\|b\|\|h\|$, we have $|\alpha(a, b)|_{\mathcal{L}(E)} \leq \|a\|\|b\|$, which implies that α is a continuous bilinear mapping. However, $\phi'(x) = \alpha(x^{-1}, x^{-1})$. As ϕ' is a composition of continuous mappings, it is continuous. It follows that ϕ is of class C^1.

Suppose now that ϕ is of class C^k. As the bilinear mapping α is of class C^k, the mapping $x \longmapsto \alpha(x^{-1}, x^{-1})$ is of class C^k, i.e., ϕ' is of class C^k. It follows that ϕ is of class C^{k+1}. By induction, ϕ is of class C^∞. $\square$

Now suppose that E and F are Banach spaces and let us consider again the subset $\mathcal{G}(E, F)$ of $\mathcal{L}(E, F)$ composed of invertible mappings. We have already seen that this subset is open and that the mapping $\psi : u \longmapsto u^{-1}$ is continuous (Theorem 1.11). We can say more about this mapping.

Corollary 4.4. *If $\mathcal{I}(E, F)$ is not empty, then the mapping*

$$\psi : \mathcal{I}(E, F) \longrightarrow \mathcal{I}(F, E), u \longmapsto u^{-1}$$

is of class C^∞.

Proof. As in the proof of Theorem 1.11, we fix $w \in \mathcal{I}(E, F)$. We have already seen that the mapping ψ may be written $\psi = \beta \circ \phi \circ \alpha$, where α is the normed vector space isomorphism from $\mathcal{L}(E, F)$ onto $\mathcal{L}(E)$ defined by $\alpha(u) = w^{-1} \circ u$, ϕ is the inversion mapping on $\mathcal{L}(E)^\times$ and β the normed vector space isomorphism from $\mathcal{L}(E)$ onto $\mathcal{L}(F, E)$ defined by $\beta(v) = v \circ w^{-1}$. From the theorem, ϕ is of class C^∞. Also β and α are continuous linear mappings and so are of class C^∞. It follows that ψ is of class C^∞. $\square$

Suppose now that E and F are normed vector spaces, $O \subset E$ and $U \subset F$ open sets and $f : O \longrightarrow U$ a diffeomorphism, i.e., f is bijective and f and f^{-1} are both differentiable. If f and f^{-1} are of class C^k, then we say that f is a C^k-*diffeomorphism* . If $f : O \longrightarrow U$ is a diffeomorphism, then

$$f^{-1} \circ f = \mathrm{id}_O \qquad \text{and} \qquad f \circ f^{-1} = \mathrm{id}_U,$$

therefore, for $x \in O$ and $y \in U$, we have

$$(f^{-1})'(f(x)) \circ f'(x) = \mathrm{id}_E \qquad \text{and} \qquad f'(f^{-1}(y)) \circ (f^{-1})'(y) = \mathrm{id}_F.$$

With $y = f(x)$, we obtain

$$f'(x) \circ (f^{-1})'(f(x)) = \mathrm{id}_F.$$

It follows that $f'(x)$ is invertible and that

$$(f'(x))^{-1} = (f^{-1})'(f(x)).$$

Now let us suppose that E and F are Banach spaces. If $y \in U$, then

$$(f^{-1})'(y) = \left(f'(f^{-1}(y)) \right)^{-1}.$$

Therefore we have the composition $(f^{-1})' = \psi \circ f' \circ f^{-1}$. If f is of class C^1, then all three mappings are continuous and it follows that f is a C^1-diffeomorphism. We can generalize this result.

Theorem 4.12. *Let O (resp. U) be an open subset of a Banach spaces E (resp. F) and $f : O \longrightarrow U$ a diffeomorphism. If f is of class C^k, then f^{-1} is also of class C^k, i.e., f is a C^k-diffeomorphism.*

Proof. We will prove the result by induction on k. For $k = 1$ we have already established the result. Now suppose that the theorem is true for k and consider the case $k + 1$. As f is class C^{k+1}, f is also of class C^k, so by the induction hypothesis f^{-1} is of class C^k. Also, f' is of class C^k, because f is of class C^{k+1}. As ψ is smooth, the composition $(f^{-1})' = \psi \circ f' \circ f^{-1}$ is of class C^k. This proves that f^{-1} is of class C^{k+1}. $\square$

Remark. Suppose that f is a differentiable mapping from an open subset O of a normed vector space E with image in a normed vector space F. If $\dim E = n < \infty$ and $\dim F = m < \infty$ and $n \neq m$, then f cannot be a diffeomorphism onto its image, because $f'(x)$ is a linear isomorphism for any point $x \in O$. This is also the case if one of the normed vector spaces is finite-dimensional and the other not.

Chapter 5
Taylor Theorems and Applications

In elementary calculus we learn certain polynomial approximations of a real-valued function in the neighbourhood of a point. These depend on the degree of differentiability of the function and, given certain conditions, it is possible to bound or estimate the error. The aim of this chapter is to generalize these approximations to mappings between normed vector spaces. In the case where the function concerned is a real-valued function defined on an open subset of $\mathbb{R}^n$, we obtain a polynomial approximation, with the polynomial being in several variables.

5.1 Taylor Formulas

We will begin with some notation.

- Let E and F be normed vector spaces, O an open subset of E containing 0 and g a mapping from O into F such that $g(0) = 0$. If there exists a mapping ϵ, defined on a neighbourhood of $0 \in E$ and with image in F, such that $\lim_{h \to 0} \epsilon(h) = 0$ and

$$g(h) = \|h\|_E^k \epsilon(h),$$

  then we will write $g(h) = o(\|h\|_E^k)$, or $o(\|h\|^k)$ when the norm is understood. If $k = 1$, then $o(\|h\|) = o(h)$.
- If E is a normed vector space and h is a vector in E, then we will write h^k for the vector $(h, \ldots, h) \in E^k$.

Lemma 5.1. *Let E and F be normed vector spaces, $\phi : E^k \longrightarrow F$ continuous k-linear and symmetric and $\Phi : E \longrightarrow F$ defined by $\Phi(x) = \phi(x^k)$. Then Φ is differentiable and*

$$\Phi'(x)h = k\phi(x^{k-1}, h)$$

for $x, h \in E$.

R. Coleman, *Calculus on Normed Vector Spaces*, Universitext,
DOI 10.1007/978-1-4614-3894-6_5, © Springer Science+Business Media New York 2012

Proof. We have

$$\Phi(x + h) = \phi(x + h, \ldots, x + h)$$
$$= \phi(x^k) + k\phi(x^{k-1}, h) + \text{terms of the form } \phi(x^p, h^q),$$

with $p + q = k$ and $q \geq 2$. The mapping $h \longmapsto k\phi(x^{k-1}, h)$ is linear and continuous; also,

$$\|\phi(x^p, h^q)\|_F \leq |\phi|_{\mathcal{L}(E^k; F)} \|x\|_E^p \|h\|_E^q.$$

The result now follows. $\square$

Theorem 5.1 (Taylor's formula, asymptotic form). *Let E and F be normed vector spaces, O an open subset of E and $a \in O$. If $f : O \longrightarrow F$ is $(k - 1)$-differentiable and $f^{(k)}(a)$ exists, then for x sufficiently small*

$$f(a + x) = f(a) + f^{(1)}(a)x + \frac{1}{2}f^{(2)}(a)(x^2) + \cdots + \frac{1}{k!}f^{(k)}(a)(x^k) + o(\|x\|^k).$$

Proof. We will prove this result by induction on k. First, by the definition of the differential, it is true for $k = 1$. We now suppose that it is true up to order $k - 1$ and consider the case k. We set

$$\phi(x) = f(a + x) - f(a) - f'(a)x - \frac{1}{2}f^{(2)}(a)(x^2) - \cdots - \frac{1}{k!}f^{(k)}(a)(x^k).$$

Using Lemma 5.1 we obtain

$$\phi'(x)h = f'(a + x)h - f'(a)h - f^{(2)}(a)(x, h) - \cdots - \frac{1}{(k - 1)!}f^{(k)}(a)(x^{k-1}, h).$$

By hypothesis, for the mapping $f' : O \longrightarrow \mathcal{L}(E, F)$ we can write

$$f'(a + x) = f'(a) + (f')'(a)x + \frac{1}{2}(f')^{(2)}(a)(x^2) + \cdots$$
$$+ \frac{1}{(k - 1)!}(f')^{(k-1)}(a)(x^{k-1}) + o(\|x\|^{k-1}),$$

therefore

$$f'(a + x)h = f'(a)h + f^{(2)}(a)(x, h) + \frac{1}{2}f^{(3)}(a)(x^2, h) + \cdots$$
$$+ \frac{1}{(k - 1)!}f^{(k)}(a)(x^{k-1}, h) + o(\|x\|^{k-1})h.$$

Hence $\phi'(x)h = o(\|x\|^{k-1})h$ and so $\phi'(x) = o(\|x\|^{k-1})$. Let us fix $\epsilon > 0$. From what we have just seen, there exists $\delta > 0$ such that $|\phi'(x)|_{\mathcal{L}(E,F)} < \epsilon\|x\|_E^{k-1}$ if

$\|x\|_E < \delta$. From Corollary 3.2 we have

$$\|\phi(x)\|_F = \|\phi(x) - \phi(0)\|_F \leq \epsilon \|x\|_E^k,$$

or

$$\left\| f(a+x) - f(a) - f'(a)x - \frac{1}{2}f^{(2)}(a)(x^2) - \cdots - \frac{1}{k!}f^{(k)}(a)(x^k) \right\|_F \leq \epsilon \|x\|_E^k.$$

It follows that

$$f(a+x) = f(a) + f^{(1)}(a)x + \frac{1}{2}f^{(2)}(a)(x^2) + \cdots + \frac{1}{k!}f^{(k)}(a)(x^k) + o(\|x\|^k).$$

Hence the result is true for k. This ends the proof.   □

Remark. If $E = \mathbb{R}^n$ and $F = \mathbb{R}$, then a consideration of the expression for $f^{(s)}(a)$ shows that $f^{(s)}(a)(x^s)$ is a homogeneous polynomial in the variables $x_1, \ldots, x_1$. Hence f can be approximated in a neighbourhood of a by a polynomial $P_{f,a,k}$ defined on $\mathbb{R}^n$. However, with the above result we have no way of estimating the error committed in taking $P_{f,a,k}(x)$ for $f(a+x)$.

Let E and F be normed vector spaces, O an open subset of E and $f : O \longrightarrow F$ k-differentiable. If we fix $h_1, \ldots, h_k \in E$, then we may define a mapping ϕ from O into F by

$$\phi(x) = f^{(k)}(x)(h_1, \ldots, h_k).$$

Proposition 5.1. *If $f^{(k+1)}(a)$ exists, then $\phi'(a)$ exists and*

$$\phi'(a)h = f^{(k+1)}(a)(h, h_1, \ldots, h_k).$$

Proof. Let α be the mapping from $\mathcal{L}(E^k; F)$ into F defined by

$$\alpha(\psi) = \psi(h_1, \ldots, h_k).$$

Then α is linear and continuous; also, $\phi = \alpha \circ f^{(k)}$. Hence

$$\phi'(a) = \alpha'(f^{(k)}(a)) \circ f^{(k)'}(a) = \alpha \circ f^{(k)'}(a).$$

Therefore

$$\phi'(a)h = \alpha(f^{(k)'}(a)h) = f^{(k)'}(a)h(h_1, \ldots, h_k) = f^{(k+1)}(a)(h, h_1, \ldots, h_k).$$

This ends the proof.   □

Lemma 5.2. *Let E and F be normed vector spaces, O an open subset of E and $f : O \longrightarrow F$ a $(k+1)$-differentiable mapping. Suppose that $a \in O$ and $h \in E$ are* *such that the segment* $[a, a + h] \in O.$ *Then the mapping*

$$\phi : [0, 1] \longrightarrow F, t \longmapsto f(a + th) + \sum_{i=1}^{k} \frac{(1-t)^i}{i!} f^{(i)}(a + th)(h^i)$$

is continuous on $[0, 1]$ *and differentiable on* $(0, 1)$ *with*

$$\frac{d\phi}{dt}(t) = \frac{(1-t)^k}{k!} f^{(k+1)}(a + th)(h^{k+1}).$$

Proof. There is no difficulty in seeing that ϕ is continuous on $[0, 1]$. If, for $i = 1, \ldots, k$, we set

$$\phi_i(x) = f^{(i)}(x)(h^i),$$

then from Proposition 5.1 ϕ_i is differentiable and

$$\frac{d}{dt} f^{(i)}(a + th)(h^i) = \frac{d}{dt}\phi_i(a + th) = \phi_i'(a + th)h = f^{(i+1)}(a + th)(h^{i+1}).$$

Applying Proposition 3.5 we obtain

$$\frac{d}{dt} \frac{(1-t)^i}{i!} f^{(i)}(a + th)(h^i) = \frac{(1-t)^i}{i!} f^{(i+1)}(a + th)(h^{i+1})$$
$$-\frac{(1-t)^{i-1}}{(i-1)!} f^{(i)}(a + th)(h^i).$$

Hence the result.   □

Theorem 5.2 (Taylor's formula, Lagrange's remainder). *Let E and F be normed vector spaces, O an open subset of E and $f : O \longrightarrow F$ a $(k + 1)$-differentiable mapping. Suppose that $a \in O$ and that $x \in E$ is such that the segment $[a, a + x]$ is contained in O. Then*

$$f(a + x) = f(a) + f^{(1)}(a)x + \frac{1}{2} f^{(2)}(a)(x^2) + \cdots + \frac{1}{k!} f^{(k)}(a)(x^k) + \mathcal{R}(a, x),$$

where

$$\|\mathcal{R}(a, x)\|_F \leq \frac{\|x\|_E^{k+1}}{(k+1)!} \sup_{0 \leq \lambda \leq 1} |f^{(k+1)}(a + \lambda x)|_{\mathcal{L}(E^{k+1}; F)}.$$

Proof. If

$$\sup_{0 \leq \lambda \leq 1} |f^{(k+1)}(a + \lambda x)|_{\mathcal{L}(E^{k+1}; F)} = \infty,$$

then there is nothing to prove, so let us suppose that this is not the case. Let ϕ be defined as in the preceding lemma. Then

$$\|\dot{\phi}(t)\|_F \leq \frac{(1-t)^k}{k!} \sup_{0 \leq \lambda \leq 1} |f^{(k+1)}(a + \lambda x)|_{\mathcal{L}(E^{k+1};F)} \|x\|_E^{k+1} = \frac{(1-t)^k}{k!} C.$$

If we set

$$\psi(t) = -\frac{(1-t)^{k+1}}{(k+1)!} C,$$

then

$$\dot{\psi}(t) = \frac{(1-t)^k}{k!} C.$$

Therefore, from Theorem 3.4,

$$\|\phi(1) - \phi(0)\|_F \leq \psi(1) - \psi(0) = \frac{C}{(k+1)!}.$$

Observing that

$$\phi(1) - \phi(0) = f(a + x) - f(a) - \sum_{i=1}^{k} \frac{1}{k!} f^{(k)}(a)(x^k),$$

we obtain the result.   □

Remark. If we know C, then we can set a bound on the remainder $\mathcal{R}(a, x)$. This is not possible with the previous Taylor formula (Theorem 5.1).

In the case where $F = \mathbb{R}$ we may obtain a particular expression for $\mathcal{R}(a, x)$ which reminds us of the remainder in the classical Taylor formula for a real-valued function defined on a compact interval of $\mathbb{R}$.

Theorem 5.3. *Let E be a normed vector space, O an open subset of E and f : $O \longrightarrow \mathbb{R}$ a $(k + 1)$-differentiable function. Suppose that $a \in O$ and that $x \in E$ is such that the segment $[a, a + x]$ is contained in O. Then there is a real number $\theta \in (0, 1)$ such that*

$$f(a + x) = f(a) + f^{(1)}(a)x + \frac{1}{2} f^{(2)}(a)(x^2) + \cdots$$

$$+ \frac{1}{k!} f^{(k)}(a)(x^k) + \frac{1}{(k+1)!} f^{(k+1)}(a + \theta x)(x^{k+1}).$$

Proof. If we set $g(t) = f(a + tx)$, then g has continuous derivatives up to order k on $[0, 1]$ and a $(k + 1)$th derivative on $(0, 1)$. An induction argument, with the help of Proposition 5.1, shows that

$$\frac{d^i g}{dt^i}(t) = f^{(i)}(a + tx)(x^i).$$

From Taylor's formula for a function defined on a compact interval of $\mathbb{R}$, we know that there is a real number $\theta \in (0, 1)$, such that

$$g(1) = g(0) + \sum_{i=1}^{k} \frac{1}{i!} \frac{d^i g}{dt^i}(0) + \frac{1}{(k+1)!} \frac{d^{k+1} g}{dt^{k+1}}(\theta),$$

or

$$f(a + x) = f(a) + \sum_{i=1}^{k} \frac{1}{i!} f^{(i)}(a)(x^i) + \frac{1}{(k+1)!} f^{(k+1)}(a + \theta x)(x^{k+1}).$$

This ends the proof. $\square$

In Theorem 5.2 we obtain a bound on the remainder $\mathcal{R}(a, x)$. We do not however determine its value. If we strengthen the hypotheses slightly, then we do obtain a precise measure of the value of $\mathcal{R}(a, x)$.

Theorem 5.4 (Taylor's formula, integral remainder). *Let E and F be normed vector spaces with F complete, O an open subset of E and $f : O \longrightarrow F$ of class C^{k+1}. If $a \in O$ and $x \in E$ is such that the segment $[a, a + x]$ is contained in O, then*

$$f(a + x) = f(a) + f^{(1)}(a)x + \frac{1}{2} f^{(2)}(a)(x^2) + \cdots$$

$$+ \frac{1}{k!} f^{(k)}(a)(x^k) + \int_0^1 \frac{(1-t)^k}{k!} f^{(k+1)}(a + tx)(x^{k+1}) dt.$$

Proof. Let ϕ be the mapping defined in Lemma 5.2. As f is of class C^{k+1}, ϕ is of class C^1 on an open interval containing $[0, 1]$. Using the fundamental theorem of calculus, we have

$$\phi(1) = \phi(0) + \int_0^1 \dot{\phi}(t) dt$$

or

$$f(a + x) = f(a) + f^{(1)}(a)x + \frac{1}{2} f^{(2)}(a)(x^2) + \cdots$$

$$+ \frac{1}{k!} f^{(k)}(a)(x^k) + \int_0^1 \frac{(1-t)^k}{k!} f^{(k+1)}(a + tx)(x^{k+1}) dt,$$

which is the result we were looking for. $\square$

5.2 Asymptotic Developments

Let E and F be normed vector spaces, O an open subset of E and f a mapping from O into F. We say that f has an asymptotic development of order k at a point $a \in O$ if there are symmetric continuous i-linear mappings A_i, for $i = 1, \ldots, k$, such that for small values of x we have

$$f(a + x) = f(a) + A_1 x + \frac{1}{2} A_2(x^2) + \cdots + \frac{1}{k!} A_k(x^k) + o(\|x\|^k).$$

From Theorem 5.1, if f is k-differentiable at a, then f has an asymptotic development of order k at a. By definition, if f has an asymptotic development of order 1 at a, then f is differentiable at a; however, f may have an asymptotic development of order $k > 1$ without being k-differentiable. Here is an example. Let $f : \mathbb{R} \longrightarrow \mathbb{R}$ be defined by

$$f(x) = \begin{cases} x^3 \sin \frac{1}{x} & x \neq 0 \\ 0 & x = 0 \end{cases}.$$

For x close to 0 we can write

$$f(x) = x^2 \left(x \sin \frac{1}{x} \right) = x^2 \epsilon(x),$$

where $\lim_{x \to 0} \epsilon(x) = 0$. Therefore f has an asymptotic development of order 2 at 0. Also,

$$\dot{f}(x) = \begin{cases} 3x^2 \sin \frac{1}{x} - x \cos \frac{1}{x} & x \neq 0 \\ 0 & x = 0 \end{cases}$$

and so

$$\frac{\dot{f}(x) - \dot{f}(0)}{x} = 3x \sin \frac{1}{x} - \cos \frac{1}{x},$$

which has no limit at 0. Therefore $\ddot{f}(0)$ does not exist and it follows that $f^{(2)}(0)$ does not exist.

In the appendix to this chapter we define and briefly study homogeneous polynomials. We use a result from this appendix in the proof of the next theorem.

Theorem 5.5. *Let E and F be normed vector spaces, O an open subset of E and f a mapping from O into F. If f has an asymptotic development at $a \in E$ of order k, then this development is unique.*

Proof. Suppose that

$$f(a + x) - f(a) = \sum_{i=1}^{k} \frac{1}{i!} A_i(x^i) + o(\|x\|^k) = \sum_{i=1}^{k} \frac{1}{i!} B_i(x^i) + o(\|x\|^k).$$

Then

$$0 = \sum_{i=1}^{k} \frac{1}{i!} C_i(x^i) + o(\|x\|^k),$$

where $C_i = A_i - B_i$. We will prove by induction on k that if this condition applies, then $C_i = 0$ for all i. If $k = 1$, then the uniqueness of the differential implies that $C_1 = 0$. Suppose now that the result is true up to order k and consider the case $k + 1$. Then we have

$$0 = \sum_{i=1}^{k+1} \frac{1}{i!} C_i(x^i) + o(\|x\|^{k+1}) = \sum_{i=1}^{k} \frac{1}{i!} C_i(x^i) + o(\|x\|^k).$$

By hypothesis $C_1 = \cdots = C_k = 0$, therefore

$$0 = \frac{1}{(k+1)!} C_{k+1}(x^{k+1}) + o(\|x\|^{k+1}).$$

If $\|x\| = 1$, then we may write

$$\left\| \frac{1}{(k+1)!} C_{k+1}((rx)^{k+1}) \right\| = \epsilon(r) r^{k+1},$$

where $\lim_{r \to 0} \epsilon(r) = 0$. On dividing by r^{k+1} we obtain $C_{k+1}(x^{k+1}) = 0$. As $C_{k+1}(x^{k+1}) = 0$ for all x of norm 1, it is so for all x. If we now apply Corollary 5.2 (see appendix), then we see that $C_{k+1} = 0$. By induction the result is true for all k.   □

Corollary 5.1. *Let E and F be normed vector spaces, O an open subset of E and f a mapping from O into F. If f has a kth differential at $a \in O$ and*

$$f(a + x) = f(a) + \sum_{i=1}^{k} \frac{1}{i!} A_i(x^i) + o(\|x\|^k),$$

then $A_i = f^{(i)}(a)$ for all i.

Proof. It is sufficient to apply Theorem 5.1 and the theorem we have just proved.□

Remark. This result may be useful in calculating higher differentials or derivatives. It may well be easier to determine an asymptotic development than to calculate a higher differential or derivative directly. This is particularly the case when handling compositions of mappings.

5.3   Extrema: Second-Order Conditions

We have previously seen that a local extremum of a differentiable function is always a critical point. We now suppose that the function is 2-differentiable at the critical point. We will concentrate on local minima; analogous results for local maxima can be easily obtained by slightly modifying the arguments.

Proposition 5.2. *Let O be an open subset of a normed vector space E, $a \in O$ and f a real-valued function defined on O having a second differential at a. If a is a local minimum, then for $h \in E$*

$$f^{(2)}(a)(h, h) \geq 0.$$

Proof. If $h = 0$, then $f^{(2)}(a)(h, h) = 0$. If $h \neq 0$, then there is an $\epsilon > 0$ such that, if $|t| < \epsilon$, then $f(a + th) \geq f(a)$. Using Theorem 5.1 and the fact that a is a critical point, we have

$$0 \leq f(a + th) - f(x) = f^{(1)}(a)th + \frac{1}{2}f^{(2)}(a)(th, th) + o(t^2 \|h\|_E^2)$$

$$= \frac{t^2}{2}f^{(2)}(a)(h, h) + o(t^2 \|h\|_E^2).$$

For $t \neq 0$, we obtain

$$0 \leq f^{(2)}(a)(h, h) + \frac{2}{t^2}o(t^2 \|h\|_E^2).$$

As $\lim_{t \to 0} \frac{o(t^2 \|h\|_E^2)}{t^2} = 0$, we have $f^{(2)}(a)(h, h) \geq 0$. □

Remark. It is sufficient to study the function $f : \mathbb{R} \longrightarrow \mathbb{R}, x \longmapsto x^3$ to see that the converse of this proposition is false.

The above result gives a necessary condition for a point to be a minimum. We now give two sufficient conditions, which however are not necessary.

Proposition 5.3. *Let O be an open subset of a normed vector space E, f a real-valued 2-differentiable function defined on O and $a \in O$ a critical point of f. If there is an open ball B centred on a such that*

$$f^{(2)}(x)(h, h) \geq 0$$

for $x \in B$ and $h \in E$, then a is a local minimum.

Proof. Let $a + h \in B$. From Theorem 5.3 and using the fact that a is a critical point, we know that there is a $\theta \in (0, 1)$ such that

$$f(a + h) = f(a) + \frac{1}{2}f^{(2)}(a + \theta h)(h, h).$$

As $a + \theta h \in B$, we have $f^{(2)}(a + \theta h)(h, h) \geq 0$ and so $f(a + h) \geq f(a)$. □

Exercise 5.1. By studying the function $f : \mathbb{R}^2 \longrightarrow \mathbb{R}, (x, y) \longmapsto x^2 y^2$, show that the converse of the above proposition is false.

Theorem 5.6. *Let O be an open subset of a normed vector space E, $a \in O$ and f a real-valued function defined on O having a second differential at a. If a is a critical point and there is an $\alpha > 0$ such that for $h \in E$*

$$f^{(2)}(a)(h, h) \geq \alpha \|h\|_E^2,$$

then a is a strict local minimum of f.

Proof. From Theorem 5.1 and taking into account that a is a critical point, for h sufficiently small we may write

$$f(a + h) - f(a) = \frac{1}{2} f^{(2)}(a)(h, h) + \epsilon(h) \|h\|^2 \geq \frac{1}{2} (\alpha + 2\epsilon(h)) \|h\|^2,$$

where $\lim_{h \to 0} \epsilon(h) = 0$. Let B be an open ball in E centred on 0 such that $|\epsilon(h)| < \frac{\alpha}{4}$ when $h \in B$. Then $\alpha + 2\epsilon(h) > \frac{\alpha}{2}$. If $h \neq 0$, then

$$f(a + h) - f(a) > \frac{\alpha}{4} \|h\|^2 > 0,$$

therefore a is a strict local minimum.   □

Exercise 5.2. By studying the function $f : \mathbb{R} \longrightarrow \mathbb{R}, x \longmapsto x^4$, show that the converse of the above theorem is false.

As above suppose that a is a critical point of f and that f is 2-differentiable at a. Let us consider $f^{[2]}(a) \in \mathcal{L}_2(E, \mathbb{R}) = \mathcal{L}(E, E^*)$, where as usual E^* is the dual $\mathcal{L}(E, \mathbb{R})$ of E. If the mapping $f^{[2]}(a)$ is an isomorphism and has a continuous inverse, then we say that a is *non-degenerate*. Notice that, if a is a non-degenerate critical point, then the norm of $f^{[2]}(a)$ is nonzero and hence so is that of $f^{(2)}(a)$.

Example. Let f be a real-valued function defined on an open subset O of $\mathbb{R}^n$ and suppose that f is 2-differentiable at a critical point $a \in O$. If H is the Hessian matrix of f at a, then for any $x \in \mathbb{R}^n$ $f^{[2]}(a)x$ is the mapping

$$l_x : \mathbb{R}^n \longrightarrow \mathbb{R}, h \longmapsto x^t H h.$$

Now $l_x = 0$ if and only if $x^t H h = 0$ for all $h \in \mathbb{R}^n$, which is equivalent to saying that $Hx = 0$, because H is symmetric. It follows that $f^{[2]}(a)$ is injective (and so bijective) if and only if H is invertible. In other words, a is non-degenerate if and only if the Hessian matrix H of f at a is invertible.

Theorem 5.7. *Let f be a real-valued function defined on an open subset of a normed vector space E. If a is a non-degenerate critical point of f and $f^{(2)}(a)(h, h) \geq 0$ for all $h \in E$, then there exists a real number $\alpha > 0$, such* *that, for $h \in E$,*

$$f^{(2)}(a)(h, h) \geq \alpha \|h\|_E$$

and therefore a is a local minimum.

Proof. To simplify the notation, let us set $L = f^{[2]}(a)$, M the norm of L^{-1} and $B = f^{(2)}(a)$. First we have $\|h\|_E \leq M |L(h)|_{E*}$. Also,

$$|L(h)|_{E*} = \sup_{\|k\|_E = 1} |L(h)k|.$$

If $h \neq 0$, then $L(h) \neq 0$ and there exists k such that $\|k\|_E = 1$ and $|L(h)k| \geq \frac{1}{2}|L(h)|_{E*}$. It follows that, for any $h \in E$, there is a $k \in E$ such that $\|k\|_E = 1$ and

$$2|L(h)k| \geq |L(h)|_{E*},$$

hence

$$\|h\|_E \leq 2M |L(h)k|.$$

We also have

$$0 \leq B(h + tk, h + tk) = t^2 B(k, k) + 2t B(h, k) + B(h, h)$$

for $t \in \mathbb{R}$. The right-hand side of this expression is a second degree polynomial in t. As it is positive for all t, its discriminant is negative, hence

$$B(h, k)^2 \leq B(k, k)B(h, h) \leq |B|_{\mathcal{L}(E^2;\mathbb{R})} B(h, h).$$

Therefore

$$\|h\|_E^2 \leq 4M^2 |B|_{\mathcal{L}(E^2;\mathbb{R})} B(h, h).$$

As $|B|_{\mathcal{L}(E^2;\mathbb{R})} \neq 0$, we can write

$$f^{(2)}(a)(h, h) = B(h, h) \geq \alpha \|h\|_E^2,$$

where $\alpha = (4M^2 |B|_{\mathcal{L}(E^2;\mathbb{R})})^{-1}$. $\qquad\square$

Let us return to the case where $E = \mathbb{R}^n$. However, first we will recall some elementary algebra. Let $M \in \mathcal{M}_n(\mathbb{R})$ be symmetric and Q the quadratic form defined by M, i.e., $Q(x) = x^t M x$ for $x \in \mathbb{R}^n$. We say that Q is definite if $\det M \neq 0$. As M is symmetric, all its eigenvalues are real and so Q is definite if and only if the eigenvalues of M are either positive or negative real numbers. If all the eigenvalues are positive (resp. negative), then we say that Q is positive (resp. negative) definite. A basis (e_i) of $\mathbb{R}^n$ is said to be orthonormal if the dot product $e_i \cdot e_j$ has the value 1 if $i = j$ and 0 otherwise. For any symmetric matrix M there is an orthonormal basis (e_i) of $\mathbb{R}^n$ composed of eigenvectors of M.

Lemma 5.3. *Let $M \in \mathcal{M}_n(\mathbb{R})$ be symmetric and Q the quadratic form defined by M. Suppose that Q is definite and let S be the unit sphere in $\mathbb{R}^n$ for the norm $\| \cdot \|_2$. If Q is positive (resp. negative) definite, then there is an $m > 0$ (resp. $m < 0$) such that $Q(x) \geq m$ (resp. $Q(x) \leq m$) for all $x \in S$. If Q is neither positive nor negative definite, then there exist $x, y \in \mathbb{R}^n \setminus \{0\}$ such that $Q(x) = 1$ and $Q(y) = -1$.*

Proof. Let (e_i) be an orthonormal basis of $\mathbb{R}^n$ composed of eigenvectors of M and (λ_i) the corresponding eigenvalues. If $x = \sum_{i=1}^{n} x_i e_i$, then $Q(x) = \sum_{i=1}^{n} \lambda_i x_i^2$. If all the eigenvalues are positive (resp. negative) and $x \in S$, then $Q(x) \geq \min \lambda_i > 0$ (resp. $Q(x) \leq \max \lambda_i < 0$). Suppose now that there are eigenvalues $\lambda_i > 0$ and $\lambda_j < 0$. If we set $x = \frac{e_i}{\sqrt{\lambda_i}}$ and $y = \frac{e_j}{\sqrt{-\lambda_j}}$, then $Q(x) = 1$ and $Q(y) = -1$. $\square$

Theorem 5.8. *Let O be an open subset of $\mathbb{R}^n$, f a real-valued 2-differentiable function defined on O, with a non-degenerate critical point a, and Q the quadratic form defined by the Hessian matrix $H = \mathcal{H}_f(a)$. We have:*

- *if Q is positive definite, then f has a strict local minimum at a;*
- *if Q is negative definite, then f has a strict local maximum at a;*
- *if Q is neither positive nor negative definite, then f does not have a local extremum at a.*

Proof. As a is non-degenerate, Q is definite. Also, $Q(x) = f^{(2)}(a)(x, x)$. Suppose first that Q is positive definite. From the above lemma, there is an $m > 0$ such that $Q(x) \geq m$ for all $x \in S$, where S is the unit sphere for the norm $\| \cdot \|_2$. This implies that $Q(x) \geq m \|x\|_2^2$ for all $x \in \mathbb{R}^n$. From Theorem 5.6 a is a strict local minimum.

If Q is negative definite, then there is an $m < 0$ such that $Q(x) \leq m$ for all $x \in S$, which implies that $-Q(x) \geq -m$ for all $x \in S$. However, $-Q$ is the quadratic form defined by the Hessian matrix of $-f$ at a. Therefore $-f$ has a strict local minimum at a and so f has a strict local maximum.

Suppose now that Q is neither positive nor negative definite. From the lemma there exist $x, y \in \mathbb{R}^n \setminus \{0\}$ such that $Q(x) = 1$ and $Q(y) = -1$. From Proposition 5.2, because $Q(y) = -1$, a cannot be a local minimum. Because $-Q(x) = -1$, $-f$ does not have a local minimum at a, hence f does not have a local maximum at a. $\square$

Remark. A non-degenerate critical point which is neither a local minimum nor a local maximum is called a *saddle point*.

Example. Let f be the real-valued function defined on $\mathbb{R}^3$ by $f(x, y, z) = z^2(1 + xy) + xy$. Then

$$\frac{\partial f}{\partial x} = yz^2 + y, \qquad \frac{\partial f}{\partial y} = xz^2 + x \qquad \text{and} \qquad \frac{\partial f}{\partial z} = 2z(1 + xy).$$

The only solution of the system $\frac{\partial f}{\partial x} = 0$, $\frac{\partial f}{\partial y} = 0$, $\frac{\partial f}{\partial z} = 0$ is the point $a = (0, 0, 0)$, so this is the unique critical point of f. The function f is 2-differentiable and

$$\mathcal{H}_f(x,y,z) = \begin{pmatrix} 0 & z^2+1 & 2yz \\ z^2+1 & 0 & 2xz \\ 2yz & 2xz & 2(1+xy) \end{pmatrix}.$$

If $H = \mathcal{H}_f(a)$, then a simple calculation shows that $\det H(a) = -2 \neq 0$ and so the associated quadratic form Q is definite. The matrix H has the eigenvalues 1, 2 and -1 and so a is a saddle point.

Exercise 5.3. Find the critical points of the real-valued function f defined by

$$f(x,y,z) = x^4 - 2x^2y + 2y^2 - 2yz + 2z^2 - 4z + 5.$$

Show that they are non-degenerate and determine the nature of each such point (local minimum, local maximum, saddle point).

Exercise 5.4. Let

$$O = \{(x_1, \dots, x_n) \in \mathbb{R}^n : x_1 > 0, \dots, x_n > 0\}$$

and $f : O \longrightarrow \mathbb{R}$ be defined by

$$f(x_1, \dots, x_n) = x_1 x_2 \cdots x_n + \alpha^{n+1} \left(\frac{1}{x_1} + \frac{1}{x_2} + \cdots + \frac{1}{x_n} \right),$$

where $\alpha \in \mathbb{R}_+^*$. Show that f has a unique critical point, which is non-degenerate, and determine its nature.

If (a,b) is a non-degenerate critical point of a 2-differentiable real-valued function f defined on an open subset of $\mathbb{R}^2$, then it is very simple to determine the nature of (a,b). Let us write

$$r = \frac{\partial^2 f}{\partial x^2}, \qquad s = \frac{\partial^2 f}{\partial x \partial y} = \frac{\partial^2 f}{\partial y \partial x} \qquad \text{and} \qquad t = \frac{\partial^2 f}{\partial y^2}.$$

Then

$$\mathcal{H}_f(x,y) = \begin{pmatrix} r & s \\ s & t \end{pmatrix}.$$

The number $rt - s^2$ is the determinant of this matrix, which is the product of the eigenvalues, and $r + t$ is the trace, which is the sum of the eigenvalues. With r, s, and t evaluated at (a,b) we obtain the following conditions:

- if $rt - s^2 > 0$ and $r > 0$, then (a,b) is a local minimum;
- if $rt - s^2 > 0$ and $r < 0$, then (a,b) is a local maximum;
- if $rt - s^2 < 0$, then (a,b) is a saddle point.

As $rt - s^2 > 0$ and $r = 0$ is impossible, this exhausts all the possibilities.

Example. Let f be the real-valued function defined on $\mathbb{R}^2$ by $f(x, y) = x^3 + 3xy^2 - 15x - 12y$. Then

$$\frac{\partial f}{\partial x} = 3x^2 + 3y^2 - 15 \quad \text{and} \quad \frac{\partial f}{\partial y} = 6xy - 12.$$

The solutions of the system $\frac{\partial f}{\partial x} = 0$, $\frac{\partial f}{\partial y} = 0$ are

$$A = (2, 1), \qquad B = (1, 2), \qquad C = (-2, -1) \quad \text{and} \quad D = (-1, -2).$$

We now calculate r, s and t:

$$r = 6x, \qquad s = 6y \quad \text{and} \quad t = 6x.$$

At B and D, $rt - s^2 = -108$, so these points are saddle points; at A, $rt - s^2 = 108$ and $r = 12$, so A is a local minimum; at C, $rt - s^2 = 108$ and $r = -12$, so C is a local maximum.

Exercise 5.5. Find the critical points of the functions f and g defined on $\mathbb{R}^2$ by

$$f(x, y) = x^3 + y^3 + 3xy \quad \text{and} \quad g(x, y) = -2x^3 + y^3 + 2x^2y - xy^2 + 3x - 3y$$

and determine the nature of each critical point.

Remark. If a critical point is degenerate, then the second differential does not help us to determine its nature. Consider the following real-valued functions defined on $\mathbb{R}$: $f : x \longmapsto x^3$, $g : x \longmapsto x^4$ and $h : x \longmapsto -x^4$. All three functions have a degenerate critical point at 0 and the same second differential. However, 0 is a local minimum of g, a local maximum of h, but neither a local minimum nor a local maximum of f.

Appendix: Homogeneous Polynomials

Suppose that E and F are vector spaces (not necessarily normed). If ϕ is a mapping from E into F and there is a k-linear mapping $\tilde{\phi}$ from E into F such that

$$\phi(x) = \tilde{\phi}(x, \ldots, x) = \tilde{\phi}(x^k)$$

for $x \in E$, then ϕ is said to be a *k-homogeneous polynomial*. The zero mapping is k-homogeneous for any k. However, for any other homogeneous polynomial, the number k is unique, because

$$\phi(\lambda x) = \lambda^k \phi(x)$$

for $\lambda \in \mathbb{R}$. We call this k the degree of ϕ. By convention we say that the zero mapping is of degree $-\infty$ and that a nonzero constant mapping from E into F is a homogeneous polynomial of degree 0.

Remark. If $F = \mathbb{R}$ and $k = 2$, then we call ϕ a *quadratic form*.

Let us write $\mathcal{H}_k(E, F)$ for the space of k-homogeneous polynomials from E into F. $\mathcal{H}_k(E, F)$ is a linear subspace of $\mathcal{F}(E, F)$, the vector space of mappings from E into F.

Proposition 5.4. *If $\phi : E \longrightarrow F$ is a k-homogeneous polynomial, then there is a symmetric k-linear mapping $\tilde{\phi}_S$ from E into F such that $\phi(x) = \tilde{\phi}_S(x^k)$.*

Proof. By definition, there is k-linear mapping $\tilde{\phi}$ from E into F such that

$$\phi(x) = \tilde{\phi}(x, \ldots, x) = \tilde{\phi}(x^k)$$

for $x \in E$. It is sufficient to set

$$\tilde{\phi}_S(x_1, \ldots, x_k) = \frac{1}{k!} \sum_{\sigma \in S_k} \tilde{\phi}(x_{\sigma(1)}, \ldots, x_{\sigma(k)}),$$

where S_k denotes the group of permutations of the set $\{1, \ldots, k\}$. $\qquad\square$

Thus we may consider homogeneous polynomials as deriving from symmetric multilinear forms.

Suppose that $E = \mathbb{R}^n$, $F = \mathbb{R}$ and (e_i) is the standard basis of $\mathbb{R}^n$. Let ϕ be a homogeneous polynomial and $\tilde{\phi}$ a k-linear form such that $\phi(x) = \tilde{\phi}(x^k)$. Then we have

$$\phi(x_1, \ldots, x_n) = \tilde{\phi}\left(\sum_{i=1}^n x_i e_i, \ldots, \sum_{i=1}^n x_i e_i\right) = \sum_{i_1, \ldots, i_k \in \mathbb{N}_n} x_{i_1} \cdots x_{i_k} \tilde{\phi}(e_{i_1}, \ldots, e_{i_k}),$$

where $\mathbb{N}_n = \{1, \ldots, n\}$. Thus ϕ is a homogeneous polynomial in the classical sense. On the other hand, if

$$\phi(x_1, \ldots, x_n) = \sum_{i_1, \ldots, i_k \in \mathbb{N}_n} c_{i_1, \ldots, i_k} x_{i_1} \ldots x_{i_k}$$

is a k-homogeneous polynomial in n variables in the classical sense and we set

$$\tilde{\phi}(e_{i_1}, \ldots, e_{i_k}) = c_{i_1, \ldots, i_k},$$

then, by extending $\tilde{\phi}$ to $(\mathbb{R}^n)^k$ in the natural way, we obtain a k-linear form such that $\phi(x) = \tilde{\phi}(x^k)$. It follows that the definition of a homogeneous polynomial which we have given above generalizes the classical definition.

As previously, let us write $L_S(E^k; F)$ for the vector space of symmetric k-linear mappings from E^k into F. The mapping

$$A : L_S(E^k; F) \longrightarrow \mathcal{H}_k(E, F), \tilde{\phi} \longmapsto \phi$$

is linear and surjective (from Proposition 5.4). Our aim is to show that this mapping is also injective. If $k = 2$, then this is easy to see: if $\tilde{\phi}$ is symmetric and $\phi = A(\tilde{\phi})$, then

$$\tilde{\phi}(x + y, x + y) = \tilde{\phi}(x, x) + \tilde{\phi}(y, y) + \tilde{\phi}(x, y) + \tilde{\phi}(y, x),$$

which implies that

$$\tilde{\phi}(x, y) = \frac{1}{2}(\phi(x + y) - \phi(x) - \phi(y)).$$

For $k > 2$ the problem is more difficult.

Let E and F be vector spaces (not necessarily normed), h an element of E and ϕ a function from E into F. We define the mapping $\Delta_h \phi : E \longrightarrow F$ by

$$\Delta_h \phi(x) = \phi(x + h) - \phi(x).$$

Δ_h is called a *difference operator* and clearly defines a linear mapping on $\mathcal{F}(E, F)$. Notice that, if ϕ is constant, then $\Delta_h \phi = 0$.

If we take two elements $h_1, h_2 \in E$, then we may define a second difference operator Δ_{h_1, h_2} by setting

$$\Delta_{h_1, h_2} \phi = \Delta_{h_1}(\Delta_{h_2} \phi).$$

We have

$$\Delta_{h_1, h_2} \phi(x) = \Delta_{h_1}(\phi(x+h_2) - \phi(x)) = \phi(x+h_2+h_1) - \phi(x+h_2) - \phi(x+h_1) + \phi(x).$$

This operator is also linear and $\Delta_{h_2,h_1} = \Delta_{h_1,h_2}$. We define higher order difference operators in an analogous way. It is not difficult to see that

$$\Delta_{h_1,\ldots,h_s,h_{s+1},\ldots,h_k}\phi = \Delta_{h_1,\ldots,h_s}(\Delta_{h_{s+1},\ldots,h_k}\phi).$$

Theorem 5.9. *If ϕ is a k-homogeneous polynomial, with $k \geq 1$, $h_1,\ldots,h_k \in E$ and $\tilde{\phi}$ is a symmetric k-linear mapping such that $\phi(x) = \tilde{\phi}(x^k)$, then for any $x \in E$*

$$\Delta_{h_1,\ldots,h_k}\phi(x) = k!\tilde{\phi}(h_1,\ldots,h_k).$$

Proof. We will prove this result by induction on k. If $k = 1$, then $\tilde{\phi} = \phi$ and the result follows. Suppose now that the result is true up to $k-1$ and consider the case k. We have

$$\Delta_{h_k}\phi(x) = \phi(x + h_k) - \phi(x) = \sum_{i=1}^{k} \binom{k}{i} \tilde{\phi}(x^{k-i}, h_k^i).$$

Using the linearity of the operator $\Delta_{h_1,\ldots,h_{k-1}}$, we obtain

$$\Delta_{h_1,\ldots,h_{k-1}}(\Delta_{h_k}\phi(x)) = \sum_{i=1}^{k} \binom{k}{i} \Delta_{h_1,\ldots,h_{k-1}}\tilde{\phi}(x^{k-i}, h_k^i).$$

The mapping $x \longmapsto \tilde{\phi}(x^{k-i}, h_k^i)$ is a $(k - i)$-homogeneous polynomial. By hypothesis

$$\Delta_{h_1,\ldots,h_{k-1}}\tilde{\phi}(x^{k-1}, h_k) = (k - 1)!\tilde{\phi}(h_1,\ldots,h_{k-1}, h_k).$$

Also, for $i \geq 2$ we have

$$\Delta_{h_i,\ldots,h_{k-1}}\tilde{\phi}(x^{k-i}, h_k^i) = (k - i)!\tilde{\phi}(h_i,\ldots,h_{k-1}, h_k^i),$$

which implies that

$$\Delta_{h_1,\ldots,h_{k-1}}\tilde{\phi}(x^{k-i}, h_k^i) = 0.$$

Therefore

$$\Delta_{h_1,\ldots,h_{k-1}}(\Delta_{h_k}\phi(x)) = k(k - 1)!\tilde{\phi}(h_1,\ldots,h_{k-1}, h_k).$$

and so the result is true for k. This ends the proof. $\square$

Corollary 5.2. *The mapping*

$$A : L_S(E^k; F) \longrightarrow \mathcal{H}_k(E, F), \tilde{\phi} \longmapsto \phi$$

is injective.

Proof. Let us fix $x \in E$. If $A(\tilde{\phi}_1) = A(\tilde{\phi}_2) = \phi$ and $h_1, \ldots, h_k \in E$, then

$$\tilde{\phi}_1(h_1, \ldots, h_k) = \frac{1}{k!}\Delta_{h_1, \ldots, h_k}\phi(x) = \tilde{\phi}_2(h_1, \ldots, h_k),$$

therefore $\tilde{\phi}_1 = \tilde{\phi}_2$ and the injectivity follows.   □

Here we have considered the generalization of homogeneous polynomials in several variables to mappings between vector spaces. It is also possible to generalize other polynomials in several variables. This more general theory is covered in Henri Cartan's book on differential calculus [6].

Chapter 6
Hilbert Spaces

In this chapter we study Hilbert spaces, which can be considered as a certain subclass of the class of normed vector spaces. We will define some particular mappings and study their differentiability.

6.1 Basic Notions

Let E be a vector space. A symmetric bilinear form $\langle \cdot, \cdot \rangle$ defined on E is said to be *positive definite* if for all $x \in E$

$$\langle x, x \rangle \geq 0 \quad \text{and} \quad \langle x, x \rangle = 0 \iff x = 0.$$

In this case we say that $\langle \cdot, \cdot \rangle$ is an *inner product*. The pair $(E, \langle \cdot, \cdot \rangle)$ is called an *inner product space*. The dot product on $\mathbb{R}^n$ is clearly an inner product and so $\mathbb{R}^n$ with the dot product is an inner product space.

Exercise 6.1. For $M, N \in \mathcal{M}_{m,n}(\mathbb{R})$, let us set

$$\langle M, N \rangle = \text{tr}\,(N^t M).$$

Show that $\langle \cdot, \cdot \rangle$ defines an inner product on $\mathcal{M}_{m,n}(\mathbb{R})$.

Proposition 6.1. *Writing $\|x\|$ for $\sqrt{\langle x, x \rangle}$, we have:*

(a) $|\langle x, y \rangle| \leq \|x\| \|y\|$;
(b) $\|x + y\| \leq \|x\| + \|y\|$;
(c) $\|x + y\|^2 + \|x - y\|^2 = 2(\|x\|^2 + \|y\|^2)$.

Proof. (a) If we define $\phi : \mathbb{R} \longrightarrow \mathbb{R}$ by $\phi(t) = \|x + ty\|^2$, then

$$0 \leq \phi(t) = \|x\|^2 + 2t \langle x, y \rangle + t^2 \|y\|^2.$$

R. Coleman, *Calculus on Normed Vector Spaces*, Universitext,
DOI 10.1007/978-1-4614-3894-6_6, © Springer Science+Business Media New York 2012

As ϕ is always nonnegative, the discriminant of ϕ must be nonpositive, i.e.,

$$4\langle x, y\rangle^2 - 4\|x\|^2\|y\|^2 \le 0.$$

Hence the result.

(b) We have

$$\|x+y\|^2 = \|x\|^2 + 2\langle x, y\rangle + \|y\|^2 \le \|x\|^2 + 2\|x\|\|y\| + \|y\|^2 = (\|x\|+\|y\|)^2,$$

therefore

$$\|x + y\| \le \|x\| + \|y\|.$$

(c) We have

$$\|x + y\|^2 + \|x - y\|^2 = (\|x\|^2 + 2\langle x, y\rangle + \|y\|^2) + (\|x\|^2 - 2\langle x, y\rangle + \|y\|^2)$$
$$= 2(\|x\|^2 + \|y\|^2).$$

This ends the proof.   □

Remarks. 1. From (b) above we see that $\|\cdot\|$ defines a norm on E (hence the notation is justifiable).
2. The inequality (a) is called *Schwarz's inequality* . It shows that an inner product is always continuous. Thus an inner product is smooth (Exercise 4.8).
3. The equality (c) is called the *parallelogram equality* by analogy with the relationship between the lengths of the sides and diagonals of a parallelogram in the plane. If a normed vector space satisfies this equality, then it can be considered an inner product space, as we will soon see.

Exercise 6.2. Prove that $|\langle x, y\rangle| = \|x\|\|y\|$ if and only if x and y are linearly dependent. What can we say for $\|x + y\| = \|x\| + \|y\|$?

Exercise 6.3. If E is an inner product space, show that the norm derived from the inner product is smooth on $E \setminus \{0\}$ and that for points $x \in E \setminus \{0\}$ and $h \in E$

$$\|x\|'h = \frac{\langle x, h\rangle}{\|x\|}.$$

Theorem 6.1. *If $(E, \|\cdot\|)$ is a normed vector space and the norm satisfies the parallelogram equality, then there is a unique inner product $\langle\cdot, \cdot\rangle$ defined on E such that $\|x\| = \sqrt{\langle x, x\rangle}$ for all $x \in E$.*

Proof. If the norm $\|\cdot\|$ is derived from an inner product $\langle\cdot, \cdot\rangle$, then

$$\|x + y\|^2 = \|x\|^2 + 2\langle x, y\rangle + \|y\|^2 \implies \langle x, y\rangle = \frac{1}{2}(\|x + y\|^2 - \|x\|^2 - \|y\|^2).$$

This proves the uniqueness and also suggests a way of defining a possible inner product. If we set

$$\alpha(x, y) = \frac{1}{2}(\|x + y\|^2 - \|x\|^2 - \|y\|^2),$$

then $\alpha(x, y) = \alpha(y, x)$ and $\alpha(x, x) = \|x\|^2$. To prove the theorem it is sufficient to show that

$$\alpha(x + x', y) = \alpha(x, y) + \alpha(x', y),$$
$$\alpha(\lambda x, y) = \lambda\alpha(x, y)$$

for all $x, x' \in E$ and $\lambda \in \mathbb{R}$. We will do this in steps.

Step 1. Two preliminary results: for all $x, y \in E$

(a) $\alpha(x, y) = \|\frac{x+y}{2}\|^2 - \|\frac{x-y}{2}\|^2$;
(b) $\alpha(x, 2y) = 2\alpha(x, y)$.

Using the parallelogram equality we have

$$\alpha(x, y) + \alpha(x, -y) = \frac{1}{2}\|x + y\|^2 + \frac{1}{2}\|x - y\|^2 - \|x\|^2 - \|y\|^2 = 0,$$

which implies that $\alpha(x, y) = -\alpha(x, -y)$. From this equality we obtain

$$\alpha(x, y) = \frac{1}{2}(\alpha(x, y) - \alpha(x, -y)) = \left\|\frac{x + y}{2}\right\|^2 - \left\|\frac{x - y}{2}\right\|^2,$$

the preliminary result (a). We now turn to (b). Using the parallelogram equality again, we obtain the two equalities

$$\left\|\frac{x}{2} + y\right\|^2 + \left\|\frac{x}{2}\right\|^2 = 2\left(\left\|\frac{x + y}{2}\right\|^2 + \left\|\frac{y}{2}\right\|^2\right);$$

$$\left\|\frac{x}{2} - y\right\|^2 + \left\|\frac{x}{2}\right\|^2 = 2\left(\left\|\frac{x - y}{2}\right\|^2 + \left\|\frac{y}{2}\right\|^2\right).$$

And using (a) we have

$$\alpha(x, 2y) = \left\|\frac{x}{2} + y\right\|^2 - \left\|\frac{x}{2} - y\right\|^2 = 2\left(\left\|\frac{x + y}{2}\right\|^2 - \left\|\frac{x - y}{2}\right\|^2\right) = 2\alpha(x, y),$$

the preliminary result (b) Notice also that, from the symmetry of α, we have

$$\alpha(x, y) = -\alpha(-x, y) \qquad \text{and} \qquad \alpha(2x, y) = 2\alpha(x, y).$$

Step 2. For all $x, x', y \in E$

$$\alpha(x + x', y) = \alpha(x, y) + \alpha(x', y).$$

From (a) and the parallelogram equality we obtain

$$\alpha(x, y) + \alpha(x', y) = \left(\left\| \frac{x+y}{2} \right\|^2 + \left\| \frac{x'+y}{2} \right\|^2 \right) - \left(\left\| \frac{x-y}{2} \right\|^2 + \left\| \frac{x'-y}{2} \right\|^2 \right)$$

$$= \frac{1}{2} \left(\left\| \frac{x+x'}{2} + y \right\|^2 + \left\| \frac{x-x'}{2} \right\|^2 \right)$$

$$- \frac{1}{2} \left(\left\| \frac{x+x'}{2} - y \right\|^2 + \left\| \frac{x-x'}{2} \right\|^2 \right)$$

$$= \frac{1}{2} \left(\left\| \frac{x+x'}{2} + y \right\|^2 - \left\| \frac{x+x'}{2} - y \right\|^2 \right)$$

$$= \frac{1}{2} \alpha(x + x', 2y).$$

Applying (b) we obtain the result.

Step 3. For all $x, y \in E$ and $\lambda \in \mathbb{R}$

$$\alpha(\lambda x, y) = \lambda \alpha(x, y).$$

First we establish by induction the result for $\lambda = n \in \mathbb{N}$. For $n = 0$ and $n = 1$ there is nothing to prove and for $n = 2$ we have proved the result above. Suppose now that statement is true for n. Then

$$\alpha((n + 1)x, y) = \alpha(nx, y) + \alpha(x, y) = n\alpha(x, y) + \alpha(x, y) = (n + 1)\alpha(x, y).$$

Hence the result is true for $n + 1$. Thus for $n \in \mathbb{N}$

$$\alpha(nx, y) = n\alpha(x, y).$$

If $n < 0$, then we have

$$\alpha(nx, y) = -\alpha(-nx, y) = -(-n\alpha(x, y)) = n\alpha(x, y).$$

This establishes the result for $\lambda = n \in \mathbb{Z}$.
  To extend this to $\mathbb{Q}$, we first notice that, if $n \in \mathbb{Z}^*$, then

$$\alpha(x, y) = \alpha\left(n\frac{1}{n}x, y \right) = n\alpha\left(\frac{1}{n}x, y \right) \implies \alpha\left(\frac{1}{n}x, y \right) = \frac{1}{n}\alpha(x, y).$$

Now, if $(p, q) \in \mathbb{Z} \times \mathbb{Z}^*$, then

$$\alpha\left(\frac{p}{q}x, y\right) = p\alpha\left(\frac{1}{q}x, y\right) = \frac{p}{q}\alpha(x, y).$$

Thus the result is true for $\lambda \in \mathbb{Q}$.

Finally, let us take $\lambda \in \mathbb{R}$. There exists a sequence $(\lambda_n) \subset \mathbb{Q}$ such that $\lim \lambda_n = \lambda$. Then

$$\lambda\alpha(x, y) = \lim \lambda_n \alpha(x, y) = \lim \alpha(\lambda_n x, y) = \alpha(\lambda x, y),$$

because $\lim \|\lambda_n x\| = \|\lambda x\|$ and $\lim \|\lambda_n x + y\| = \|\lambda x + y\|$.

We have proved that α is an inner product. This ends the proof.   □

Example. The norm $\|\cdot\|_2$ on $\mathbb{R}^n$ is derived from an inner product. However, this is not the case for the norms $\|\cdot\|_p$ with $p \neq 2$. If

$$x = (1, 0, 0, \ldots, 0) \qquad \text{and} \qquad y = (0, 1, 0, \ldots, 0),$$

then simple calculations show that the parallelogram equality is not satisfied if $1 \leq p < \infty$, $p \neq 2$. If we replace x by

$$x' = (1, 1, 0, \ldots, 0),$$

then the parallelogram equality is not satisfied for $p = \infty$.

If E is an inner product space and complete with the norm derived from the inner product, then E is said to be a *Hilbert space*. (In this case we usually write H for E.) Thus $\mathbb{R}^n$ with the dot product is a Hilbert space. However, not all inner product spaces are Hilbert spaces as the next exercise shows.

Exercise 6.4. Let E be the space of continuous real-valued functions defined on the closed interval $[-1, 1]$. For $f, g \in E$, let $\langle f, g \rangle = \int_{-1}^{1} fg$. Show that $\langle \cdot, \cdot \rangle$ defines an inner product on E. Using the sequence of functions (f_n) defined by

$$f_n(x) = \begin{cases} 0 & -1 \leq x \leq -\frac{1}{n} \\ nx + 1 & -\frac{1}{n} < x \leq 0 \\ 1 & 0 < x \leq 1 \end{cases},$$

show that E is not a Hilbert space.

Exercise 6.5. Consider the space l^2 composed of sequences of real numbers (x_n) such that $\sum_{n=1}^{\infty} |x_n|^2 < \infty$. Show that, if $(x_n), (y_n) \in l^2$ and we set

$$\langle (x_n), (y_n) \rangle = \sum_{n=1}^{\infty} x_n y_n,$$

then $\langle \cdot, \cdot \rangle$ defines an inner product on l^2 and that l^2 with this inner product is a Hilbert space.

A Hilbert space H may be a Banach algebra. It is interesting to notice that in this case H is always a division ring. (An easy proof of this may be found in [7].) This implies that H is isomorphic to $\mathbb{R}$, $\mathbb{C}$ or $\mathbb{H}$ (see Sect. 1.7).

If x and y are nonzero vectors in the plane $\mathbb{R}^2$, based at the origin, and θ is the angle between them, then it is easy to show that $x \cdot y = |x||y|\cos\theta$, where $|x|$ (resp. $|y|$) is the length of x (resp. y). This equality is equivalent to the statement $\arccos\theta = \frac{x \cdot y}{|x||y|}$. We can generalize this. If x and y are nonzero elements of an inner product space E, then we define the angle $\theta(x, y)$ between x and y as follows:

$$\theta(x, y) = \arccos \frac{\langle x, y \rangle}{\|x\| \|y\|}.$$

The function arccos is defined on the interval $[-1, 1]$. By Schwarz's inequality $\frac{\langle x,y \rangle}{\|x\| \|y\|} \in [-1, 1]$ and so this definition makes sense.

6.2   Projections

If K is a nonempty closed subset of a normed vector space and $y \in H$, then we define the distance from y to K as follows:

$$d_K(y) = \inf_{z \in K} \|y - z\|.$$

If K is a convex subset of a Hilbert space, then this distance is realized by a unique point $P_K(y) \in K$, which is called the projection of y on K. We thus obtain a mapping P_K of H onto K. In this section we will study the projection mapping P_K and in the next the distance mapping d_K.

Theorem 6.2. *Let H be a Hilbert space and K a closed convex subset of H, which is not empty. For each $y \in H$ there exists a unique element $x \in K$ such that*

$$\|y - x\| = \inf_{z \in K} \|y - z\|. \tag{6.1}$$

Furthermore, x is characterized by the property

$$x \in K \qquad and \qquad \langle y - x, z - x \rangle \le 0 \tag{6.2}$$

for all $z \in K$.

Proof. As the proof is a little long, we will proceed by steps.

Step 1. Existence of the unique element x.

Let (z_n) be a sequence in K such that

$$\lim \|y - z_n\| = \inf_{z \in K} \|y - z\|.$$

To simplify the notation let us set $d_n = \|y - z_n\|$ and $d = \inf_{z \in K} \|y - z\|$. Applying the parallelogram equality

$$\|a + b\|^2 + \|a - b\|^2 = 2(\|a\|^2 + \|b\|^2),$$

with $a = y - z_n$ and $b = y - z_m$, we obtain

$$\left\| y - \frac{z_n + z_m}{2} \right\|^2 + \left\| \frac{z_n - z_m}{2} \right\|^2 = \frac{1}{2}(d_n^2 + d_m^2).$$

As K is convex, $\frac{z_n + z_m}{2} \in K$ and so $\|y - \frac{z_n + z_m}{2}\| \geq d$. Therefore

$$\left\| \frac{z_n - z_m}{2} \right\|^2 \leq \frac{1}{2}(d_n^2 + d_m^2) - d^2 = \frac{1}{2}(d_n^2 - d^2) + \frac{1}{2}(d_m^2 - d^2).$$

This implies that (z_n) is a Cauchy sequence. If we set $x = \lim z_n$, then $x \in K$, because K is closed. To see that $\|x - y\| = d$ it is sufficient to notice that $\|x - y\| \leq \|x - z_n\| + \|z_n - y\|$.

Step 2. Equivalence of the conditions (6.1) and (6.2).

First suppose that x satisfies (6.1). If $z \in K$ and $t \in (0, 1]$, then $(1 - t)x + tz \in K$, hence

$$\|y - x\| \leq \|y - [(1 - t)x + tz]\| = \|y - x - t(z - x)\|$$

and so

$$\|y - x\|^2 \leq \|y - x\|^2 - 2t\langle y - x, z - x \rangle + t^2 \|z - x\|^2.$$

This gives us the inequality $2\langle y - x, z - x \rangle \leq t \|z - x\|^2$. If we now let t go to 0, we obtain $\langle y - x, z - x \rangle \leq 0$. This proves that (6.1) implies (6.2).

Now suppose that x satisfies (6.2). If $z \in K$, then

$$\|y - z\|^2 = \|(y - x) - (z - x)\|^2 = \|y - x\|^2 - 2\langle y - x, z - x \rangle + \|z - x\|^2,$$

which implies that

$$\|y - x\|^2 = \|y - z\|^2 + 2\langle y - x, z - x \rangle - \|z - x\|^2 \leq \|y - z\|^2,$$

because $\langle y - x, z - x \rangle \leq 0$. We have thus shown that (6.2) implies (6.1) and we have the equivalence we were looking for.

Step 3. Uniqueness of the element x.

Suppose that x_1 and x_2 satisfy (6.2). Then for any $z \in K$ we have

$$\langle y - x_1, z - x_1 \rangle \leq 0 \qquad \text{and} \qquad \langle y - x_2, z - x_2 \rangle \leq 0.$$

As $x_1, x_2 \in K$, we can replace z by x_2 in the first expression and by x_1 in the second to obtain

$$\langle y - x_1, x_2 - x_1 \rangle \leq 0 \qquad \text{and} \qquad \langle x_2 - y, x_2 - x_1 \rangle \leq 0.$$

On adding the two inequalities, we see that $\|x_2 - x_1\|^2 \leq 0$, which implies that $x_2 = x_1$. This proves the uniqueness and so finishes the proof. $\square$

Corollary 6.1. *A closed convex subset K of a Hilbert space H contains a unique element of minimum norm.*

Proof. It is sufficient to set $y = 0$ in the *theorem*. $\square$

As we have already stated, the unique element $x = P_K(y)$ which minimises $\|y - z\|$ for $z \in K$ is called the *projection* of y onto K and the mapping $y \longmapsto P_K(y)$ from H into itself the projection (mapping) of H onto K. There is no difficulty in seeing that $P_K(y) = y$ if and only if $y \in K$. In the next proposition we will look at some properties of the projection of an element lying outside of K.

Proposition 6.2. *If $y \notin K$, then $x = P_K(y) \in \partial K$. Also, for all vectors $z \in K$, the angle between the vectors $y - x$ and $z - x$ is not less than $\frac{\pi}{2}$.*

Proof. If $x \notin \partial K$, then $x \in \text{int}\, K$ and there is an open ball $B(x, r)$ lying in K. For $t \in (0, 1)$ sufficiently small, $x + t(y - x) \in B(x, r)$ and

$$\|y - (x + t(y - x))\| = (1 - t)\|y - x\| < \|y - x\|,$$

a contradiction. Hence $x \in \partial K$.

To prove the second statement it is sufficient to notice that

$$\theta(y - x, z - x) = \arccos \frac{\langle y - x, z - x \rangle}{\|y - x\| \|z - x\|} \leq 0.$$

This ends the proof. $\square$

Exercise 6.6. Suppose that $y \notin K$, $x = P_K(y)$ and that y' belongs to the segment (x, y). Show that $P_K(y') = x$.

Example. We consider $\mathbb{R}^n$ with the dot product. The set $K = \mathbb{R}^n_+$ is closed and convex. We claim that $P_K(y)_i = \max\{y_i, 0\}$. We have

$$\langle y - x, z - x \rangle = \sum_{i=1}^{n} (y_i - x_i)(z_i - x_i).$$

Let $z \in K$ and $x_i = \max\{y_i, 0\}$. Then

$$y_i \leq 0 \Longrightarrow x_i = 0 \Longrightarrow y_i - x_i = y_i \leq 0 \qquad \text{and} \qquad z_i - x_i = z_i \geq 0.$$

Also,

$$y_i > 0 \Longrightarrow x_i = y_i \Longrightarrow y_i - x_i = 0.$$

Therefore $\langle y - x, z - x \rangle \leq 0$ and the affirmation is proved.

Exercise 6.7. Let $K = \{(x_1, x_2) \in \mathbb{R}^2 : x_1 + t x_2 \geq 0, t \in [-1, 1]\}$. Show that K is a closed convex subset of $\mathbb{R}^2$. Find the set of points (y_1, y_2) whose projection onto K is the origin.

Exercise 6.8. If K is the closed unit ball in a Hilbert space and $y \notin K$, show that the projection of y onto K is the point $\frac{y}{\|y\|}$.

If K is a closed convex subset of a normed vector space E, which is not a Hilbert space, then it is not always possible to define a projection mapping: there may be no element realizing the minimum distance or more than one element realizing this distance. Let us look at two examples.

- Let $E = C([0, 1], \mathbb{R})$ be the vector space of real-valued continuous functions defined on the interval $[0, 1]$. If we set $\|f\| = \sup_{t \in [0,1]}\{|f(t)|\}$, then $\| \cdot \|$ defines a norm on E and $(E, \| \cdot \|)$ is a Banach space. Let

$$K = \left\{ f \in E : \int_0^{\frac{1}{2}} f(t)\mathrm{d}t - \int_{\frac{1}{2}}^1 f(t)\mathrm{d}t = 1 \right\}.$$

Then K is a nonempty closed convex subset of E having no element of minimum norm. Hence we cannot define the projection of 0 on K.

- Let $E = l^1(\mathbb{R})$ be the vector space of real sequences $x = (x_n)_{n=1}^{\infty}$ such that $\sum_{n=1}^{\infty} |x_n| < \infty$. We have seen that, if we set $\|x\| = \sum_{n=1}^{\infty} |x_n|$, then $\| \cdot \|$ defines a norm on E and that $(E, \| \cdot \|)$ is a Banach space. The set

$$K = \left\{ x \in E : \sum_{n=1}^{\infty} x_n = 1 \right\},$$

is a closed convex subset of E. K cannot have a unique element of minimum norm, because there are always many elements having the same norm. Thus we cannot define the projection of 0 onto K.

We will now study the projection mapping in more detail. However, first a definition. Let E and F be normed vector spaces and $k \in \mathbb{R}_+$. We say that

$f : E \longrightarrow F$ is a k-*Lipschitz mapping* if

$$\|f(x) - f(y)\|_F \le k\|x - y\|_E$$

for all $x, y \in E$. If $k < 1$ we call f a *contraction mapping* or contraction. Clearly Lipschitz mappings are continuous.

Proposition 6.3. *The projection P_K is a 1-Lipschitz mapping.*

Proof. If $x_1 = P_K(y_1)$ and $x_2 = P_K(y_2)$, then for $z \in K$

$$\langle y_1 - x_1, z - x_1 \rangle \le 0 \qquad \text{and} \qquad \langle y_2 - x_2, z - x_2 \rangle \le 0.$$

Remembering that $x_1, x_2 \in K$, we obtain

$$\langle y_1 - x_1, x_2 - x_1 \rangle \le 0 \qquad \text{and} \qquad \langle x_2 - y_2, x_2 - x_1 \rangle \le 0.$$

Summing the two expressions and using the Schwarz inequality, we obtain

$$\|x_2 - x_1\|^2 \le \langle y_2 - y_1, x_2 - x_1 \rangle \le \|y_2 - y_1\| \|x_2 - x_1\|,$$

from which we deduce $\|x_2 - x_1\| \le \|y_2 - y_1\|$.   □

Although the projection mapping is 1-Lipschitz (and hence continuous), it is not in general differentiable on its domain. This depends on the nature of the closed convex set K. Let us return to the example given just after Exercise 6.6. The ith coordinate function P_{K_i} can be written

$$P_{K_i}(y) = \begin{cases} y_i & \text{when } y_i \ge 0, \\ 0 & \text{when } y_i < 0. \end{cases}$$

If $y_i \ne 0$, then the partial derivative $\frac{\partial P_{K_i}}{\partial y_i}$ is defined and continuous on a neighbourhood of y. However, if $y_i = 0$, then the partial derivative $\frac{\partial P_{K_i}}{\partial y_i}(y)$ is not defined. It follows that P_{K_i} is differentiable only at those points y whose coordinates are all nonzero.

Let us now turn to the particular case where K is a closed vector subspace of H. In this case we have a simpler characterization of the projection of an element and the projection mapping is everywhere differentiable.

Proposition 6.4. *If K is a closed vector subspace of a Hilbert space H and $y \in H$, then $x = P_K(y)$ can be characterized as follows:*

$$x \in K \qquad \text{and} \qquad \langle y - x, z \rangle = 0 \tag{6.3}$$

for all $z \in K$. *In addition, the projection* P_K *is a linear continuous mapping of norm not greater than* 1.

Proof. Let us fix $z \in K$. If $t \in \mathbb{R}$, then $tz \in K$ and so for $t \in \mathbb{R}$

$$\langle y - x, tz - x \rangle \leq 0.$$

This implies that

$$t\langle y - x, z \rangle \leq \langle y - x, x \rangle$$

and so $\langle y - x, z \rangle = 0$. Conversely, suppose that x satisfies the condition (6.3). Then for $z \in K$

$$\langle y - x, z - x \rangle = 0,$$

because $z - x \in K$. Therefore $x = P_K(y)$. This finishes the first part of the proof.

Let $\alpha \in \mathbb{R}$, $x = P_K(y)$ and $z \in K$; then

$$\langle y - x, z \rangle = 0 \implies \alpha \langle y - x, z \rangle = 0 \implies \langle \alpha y - \alpha x, z \rangle = 0$$

and so $P_K(\alpha y) = \alpha P_K(y)$. Suppose now that $x_1 = P_K(y_1)$ and $x_2 = P_K(y_2)$. Then

$$\langle y_1 - x_1, z \rangle = 0 \quad \text{and} \quad \langle y_2 - x_2, z \rangle = 0 \implies \langle (y_1 + y_2) - (x_1 + x_2), z \rangle = 0$$

and it follows that $P_K(y_1 + y_2) = P_K(y_1) + P_K(y_2)$. This proves that P_K is linear. In addition,

$$\|P_K(y)\| = \|P_K(y) - P_K(0)\| \leq \|y - 0\| = \|y\|$$

and so P_K is continuous with norm not greater than 1. $\square$

Corollary 6.2. *If K is a closed affine subspace of a Hilbert space H, then the projection mapping P_K is smooth on H.*

Proof. If K is a subspace of H, then the result follows directly from the fact that P_K is linear and continuous. Suppose now that $K = a + \tilde{K}$, where $\tilde{K}$ is a subspace and $a \notin \tilde{K}$; then $\tilde{K}$ is closed, because the mapping $y \longmapsto a + y$ is a homeomorphism. If $x = P_{\tilde{K}}(y)$, then

$$\|(a + y) - (a + x)\| = \|y - x\| = \inf_{z \in \tilde{K}} \|y - z\| = \inf_{z \in \tilde{K}} \|(a + y) - (a + z)\|,$$

therefore

$$P_K(a + y) = P_{\tilde{K}}(y) + a \implies P_K(y) = P_{\tilde{K}}(y - a) + a = P_{\tilde{K}}(y) - P_{\tilde{K}}(a) + a.$$

Hence $P_K = P_{\tilde{K}} - P_{\tilde{K}}(a) + a$ and it follows that P_K is smooth. $\square$

6.3   The Distance Mapping

If K is a nonempty closed subset of a normed vector space E, then, as we have seen, we define the distance of a point $y \in E$ to K by

$$d_K(y) = \inf_{z \in K} \|y - z\|.$$

Proposition 6.5. *If K is a nonempty closed subset of a normed vector space, then the mapping d_K is 1-Lipschitz and therefore continuous.*

Proof. Let $y_1, y_2 \in E$ and $z \in K$. Then

$$|\,\|y_1 - z\| - \|y_2 - z\|\,| \le \|(y_1 - z) - (y_2 - z)\| = \|y_1 - y_2\|.$$

Therefore

$$\|y_1 - z\| \le \|y_2 - z\| + \|y_1 - y_2\|$$

and we easily obtain

$$\inf_{z \in K} \|y_1 - z\| \le \inf_{z \in K} \|y_2 - z\| + \|y_1 - y_2\|.$$

Changing the roles of y_1 and y_2 we obtain

$$\inf_{z \in K} \|y_2 - z\| \le \inf_{z \in K} \|y_1 - z\| + \|y_2 - y_1\|.$$

It follows that

$$|d_K(y_1) - d_K(y_2)| \le \|y_2 - y_1\|,$$

and so d_K is 1-Lipschitz. $\qquad\square$

The mapping d_K is not in general differentiable, even on a Hilbert space. However, if we take the square of d_K, then we obtain a differentiable mapping, as we will now see.

Theorem 6.3. *If K is a nonempty closed convex subset of a Hilbert space, then the mapping d_K^2 is of class C^1.*

Proof. Let us fix $x \in E$ and write $\Delta(h) = d_K^2(x + h) - d_K^2(x)$. As $d_K^2(x + h) \le \|x + h - P_K(x)\|^2$, we have

$$-\Delta(h) \ge \|x - P_K(x)\|^2 - \|x + h - P_K(x)\|^2$$
$$= \langle x - P_K(x), x - P_K(x) \rangle - \langle x + h - P_K(x), x + h - P_K(x) \rangle$$
$$= -\|h\|^2 - 2\langle x - P_K(x), h \rangle,$$

which implies that
$$\Delta(h) - 2\langle x - P_K(x), h \rangle \leq \|h\|^2.$$

Now, $d_K^2(x) \leq \|x - P_K(x + h)\|^2$, therefore

$$\Delta(h) \geq \|x + h - P_K(x + h)\|^2 - \|x - P_K(x + h)\|^2$$
$$= \langle x + h - P_K(x + h), x + h - P_K(x + h) \rangle$$
$$- \langle x - P_K(x + h), x - P_K(x + h) \rangle$$
$$= \|h\|^2 + 2\langle x - P_K(x + h), h \rangle$$

and so

$$-\Delta(h) + 2\langle x - P_K(x), h \rangle \leq -\|h\|^2 - 2\langle x - P_K(x + h), h \rangle + 2\langle x - P_K(x), h \rangle$$
$$= -\|h\|^2 + 2\langle P_K(x + h) - P_K(x), h \rangle$$
$$\leq -\|h\|^2 + 2\|P_K(x + h) - P_K(x)\|\|h\|$$
$$\leq -\|h\|^2 + 2\|h\|^2 = \|h\|^2.$$

Hence

$$\frac{|\Delta(h) - 2\langle x - P_K(x), h \rangle|}{\|h\|} \leq \|h\|,$$

which implies that d_K^2 is differentiable and

$$(d_K^2)'(x)h = 2\langle x - P_K(x), h \rangle.$$

Also,

$$((d_K^2)'(x + u) - (d_K^2)'(x))h = 2\langle (x + u) - P_K(x + u), h \rangle - 2\langle x - P_K(x), h \rangle$$
$$= 2\langle u - P_K(x + u) + P_K(x), h \rangle,$$

therefore

$$|(d_K^2)'(x + u) - (d_K^2)'(x)|_{H^*} \leq 2\|u - P_K(x + u) + P_K(x)\|_H,$$

which implies that $(d_K^2)'$ is continuous. Thus d_K^2 is of class C^1.   □

We will see a little later that the function d_K^2 has another interesting property.

6.4   The Riesz Representation Theorem

If H is a Hilbert space, then by the Riesz representation theorem, which we will presently prove, we may associate an element of H to a continuous linear form. Before looking at the general case, let us see what happens in $\mathbb{R}^n$. If l is a linear form defined on $\mathbb{R}^n$, (e_i) its standard basis and $x = \sum_{i=1}^{n} x_i e_i$ then

$$l(x) = \sum_{i=1}^{n} x_i l(e_i) = x \cdot w,$$

where $w = (l(e_1), \ldots, l(e_n))$. If $\bar{w}$ is such that $l(x) = x \cdot \bar{w}$ for all $x \in \mathbb{R}^n$, then $x \cdot (w - \bar{w}) = 0$ for all $x \in \mathbb{R}^n$, and it follows that $w - \bar{w} = 0$. Hence the element w such that $l(x) = x \cdot w$ for all x is unique. We will now look at the general case. However, to do so, we need a new concept.

If S is a subset of an inner product space E, then we define $S^\perp$ by

$$S^\perp = \{x \in E : \langle x, s \rangle = 0 \text{ for all } s \in S\}.$$

$S^\perp$ is said to be the *orthogonal complement* of S. It is easy to see that $S^\perp$ is a closed vector subspace of E.

Exercise 6.9. Show that, if S is a closed subspace of a Hilbert space H, then S and $S^\perp$ are complementary.

Now let us turn to the Riesz representation theorem.

Theorem 6.4 (Riesz representation theorem). *Let l be a continuous linear form defined on a Hilbert space H. Then there is a unique element $a \in H$ such that*

$$l(x) = \langle x, a \rangle$$

for all $x \in H$. In addition, $|l|_{H^*} = \|a\|$.

Proof. Let $S = \operatorname{Ker} l$. If $H = S$, then $l \equiv 0$ and we can set $a = 0$. If $S \neq H$, then there exists w such that $l(w) \neq 0$. However, $w = y + z$, with $y \in S$ and $z \in S^\perp$, and so $l(z) \neq 0$. For $x \in H$ we have

$$l\left(x - \frac{l(x)}{l(z)} z\right) = l(x) - \frac{l(x)}{l(z)} l(z) = 0$$

and so $x - \frac{l(x)}{l(z)} z \in S$. This implies that $\langle x - \frac{l(x)}{l(z)} z, z \rangle = 0$, or $\langle x, z \rangle = \frac{l(x)}{l(z)} \|z\|^2$. If we set $a = \frac{l(z)}{\|z\|^2} z$, then we obtain

$$\langle x, a \rangle = l(x).$$

Suppose now that both a_1 and a_2 satisfy the condition of the theorem. Then, for all $x \in H$

$$\langle x, a_1 - a_2 \rangle = 0.$$

This implies that $a_1 - a_2 = 0$ and so the element a is unique.

Finally, we must show that $|l|_{H^*} = \|a\|$. If $l \equiv 0$, then the result is clear, so suppose that this is not the case. Then $a \neq 0$. If $\|x\| \leq 1$, then

$$|l(x)| = |\langle x, a \rangle| \leq \|x\| \|a\| \leq \|a\|$$

and so $|l|_{H^*} \leq \|a\|$. In addition, $\left\| \frac{a}{\|a\|} \right\| = 1$ and

$$\left| l \left(\frac{a}{\|a\|} \right) \right| = \left| \left\langle \frac{a}{\|a\|}, a \right\rangle \right| = \|a\|,$$

therefore $|l|_{H^*} = \|a\|$.   □

Exercise 6.10. Let S be a vector subspace of a Hilbert space H and l a continuous linear form defined on S. Show that l has a unique extension to a continuous linear form defined on H which preserves the norm of l.

Remark. If E is any normed vector space, S a vector subspace and l a continuous linear form defined on S, then there exists an extension of l to E which preserves the norm. This is proved in any standard text on functional analysis. However, the extension may not be unique. For example, if $E = l^1$ and we set

$$S = \{ x \in l^1 : x_i = 0, i \geq 2 \},$$

then S is a closed subspace of E and the mapping $l : S \longrightarrow \mathbb{R}, x \longmapsto x_1$ is a continuous linear form which can be extended in an infinite number of ways to E, all of which preserve the norm.

If f is a real-valued mapping defined on an open subset O of a Hilbert space H and is differentiable at a point $x \in O$, then $f'(x)$ is a continuous linear form and so, from Theorem 6.4, there is a unique element $a \in H$ such that

$$f'(x)h = \langle h, a \rangle$$

for all $h \in H$. We call a the *gradient* of f at x and write $\nabla f(x)$ for a. If f is differentiable on O, then we obtain a mapping ∇f from O into H to which we also give the name gradient.

Theorem 6.5. *If f has a second differential at a point $x \in O$, then ∇f is differentiable at x. If f is of class C^2 on O, then ∇f is of class C^1 on O.*

Proof. If we fix h, then the mapping $\phi_h : k \longmapsto f^{(2)}(x)(h, k)$ is a continuous linear form, so there exists $v_x(h) \in H$ such that

$$f^{(2)}(x)(h,k) = \langle v_x(h), k \rangle$$

for all $k \in H$. It is a simple matter to check that v_x is linear. In addition,

$$\|v_x(h)\|_H = |\phi_h|_{H^*} = \sup_{\|k\|_H \leq 1} |f^{(2)}(x)(h,k)| \leq |f^{(2)}(x)|_{\mathcal{L}(H^2;\mathbb{R})} \|h\|_H,$$

hence v_x is continuous and

$$|v_x|_{\mathcal{L}(H)} \leq |f^{(2)}(x)|_{\mathcal{L}(H^2;\mathbb{R})}.$$

Now,

$$f'(x+h)k = f'(x)k + f^{(2)}(x)(h,k) + o(h)k$$

can be written

$$\langle \nabla f(x+h), k \rangle = \langle \nabla f(x), k \rangle + \langle v_x(h), k \rangle + \langle \nabla o(h), k \rangle$$

and therefore

$$\nabla f(x+h) = \nabla f(x) + v_x(h) + \nabla o(h).$$

As $\|\nabla o(h)\|_H = |o(h)|_{H^*}$, we have proved that $(\nabla f)'(x) = v_x$, i.e., ∇f is differentiable at x.

Now suppose that f is of class C^2 on O. Then for $x \in O$ and u small we have

$$|v_{x+u} - v_x|_{\mathcal{L}(H)} \leq |f^{(2)}(x+u) - f^{(2)}(x)|_{\mathcal{L}(H^2;\mathbb{R})}.$$

The continuity of $f^{(2)}$ at x implies that of v at x. Therefore ∇f is of class C^1 on O. $\qquad\qquad\square$

Remark. We write $\nabla^2(x)$ for v_x. If $H = \mathbb{R}^n$ with the dot product, then $\nabla^2(x)(h) = \mathcal{H}f(x)h$ for all $h \in \mathbb{R}^n$, where $\mathcal{H}f(x)$ is the Hessian matrix of f at x.

Chapter 7
Convex Functions

Let X be a convex subset of a vector space V. We say that $f : X \longrightarrow \mathbb{R}$ is *convex* if for all $x, y \in X$ and $\lambda \in (0, 1)$ we have

$$f(\lambda x + (1 - \lambda)y) \leq \lambda f(x) + (1 - \lambda) f(y).$$

If the inequality is strict when $x \neq y$, then we say that f is *strictly convex*. In this chapter we aim to look at some properties of these functions, in particular, when E is a normed vector space. For differentiable functions we will obtain a characterization, which will enable us to generalize the concept of a convex function.

Exercise 7.1. We have seen that a norm on a vector space is a convex function. Is a norm strictly convex?

7.1 Preliminary Results

In this section we will introduce some elementary results on convex sets and convex functions.

Exercise 7.2. If C is a subset of a vector space V such that $x + y \in C$ when $x, y \in C$, and $\alpha x \in C$ when $x \in C$ and $\alpha \in \mathbb{R}_+$, then we say that C is a convex cone. Show that the convex functions defined on a convex set X form a convex cone in the vector space of real-valued functions defined on X.

Proposition 7.1. *Let X be a convex subset of a normed vector space E. Then*

(a) *$\bar{X}$ is convex;*
(b) *int X is convex.*

Proof. (a) If $x, y \in \bar{X}$, then there are sequences $(x_n), (y_n) \subset X$ which converge respectively to x and y. If $\lambda \in (0, 1)$, then the sequence $(\lambda x_n + (1 - \lambda)y_n)$

R. Coleman, *Calculus on Normed Vector Spaces*, Universitext,
DOI 10.1007/978-1-4614-3894-6_7, © Springer Science+Business Media New York 2012

converges to $\lambda x + (1-\lambda)y$. As $\lambda x_n + (1-\lambda)y_n \in X$ for all n, $\lambda x + (1-\lambda)y \in \bar{X}$. It follows that $\bar{X}$ is convex.

(b) Suppose that $x, y \in \text{int } X$. There exists $r > 0$ such that the open balls $B(x, r)$ and $B(y, r)$ are included in int X. If $\|u\| < r$ and $\lambda \in (0, 1)$, then

$$\lambda x + (1 - \lambda)y + u = \lambda(x + u) + (1 - \lambda)(y + u) \in X.$$

Therefore

$$B(\lambda x + (1 - \lambda)y, r) \subset X$$

and so $\lambda x + (1 - \lambda)y \in \text{int } X$. It follows that int X is convex. $\square$

Let E be a vector space and $x_1, \ldots, x_n \in E$. We say that $y \in E$ is a *convex combination* of the points $x_1, \ldots, x_n$ if there exist $\lambda_1, \ldots, \lambda_n \in [0, 1]$, with $\sum_{i=1}^{n} \lambda_i = 1$, such that $y = \sum_{i=1}^{n} \lambda_i x_i$. If X is a nonempty subset of E, then we define co X, the *convex hull* of X, to be the set of points $y \in E$ which are convex combinations of points in X.

Proposition 7.2. *co X is convex and is the intersection of all convex subsets containing X.*

Proof. Let us first show that co X is convex. We take $x, y \in \text{co } X$, with $x = \sum_{i=1}^{p} \lambda_i x_i$ and $y = \sum_{j=1}^{q} \mu_j y_j$, and $\theta \in [0, 1]$. Then

$$\theta x + (1 - \theta)y = \sum_{i=1}^{p} (\theta \lambda_i) x_i + \sum_{j=1}^{q} ((1 - \theta)\mu_j) y_j.$$

As the right-hand side of the expression is a convex combination of elements in X, $\theta x + (1 - \theta)y \in \text{co } X$ and it follows that co X is convex.

Now let us consider the second part of the proposition. Let Y be a convex set. We will prove by induction on p, the number of elements in a convex combination, that any convex combination of elements of Y is an element of Y. If $p = 1$ or $p = 2$, then the statement is clearly true. Suppose that the result is true up to p and let $y = \sum_{i=1}^{p+1} \lambda_i y_i$ be a convex combination of elements of Y. If $\sum_{i=1}^{p} \lambda_i = 0$, then $y = y_{p+1} \in Y$. Otherwise, if we set

$$\lambda_i' = \frac{\lambda_i}{\sum_{i=1}^{p} \lambda_i}$$

for $i = 1, \ldots, p$, then by the induction hypothesis $y' = \sum_{i=1}^{p} \lambda_i' y_i$ belongs to Y. However,

$$y = \left(\sum_{i=1}^{p} \lambda_i \right) y' + \lambda_{p+1} y_{p+1},$$

which belongs to Y, because Y is convex. So, by induction, the statement is true for any p.

Suppose now that Y is convex and that $X \subset Y$. From what we have just seen, any convex combination of elements of X lies in Y, therefore $\operatorname{co} X \subset Y$. As the intersection K of all convex sets containing X is convex and contains X, $\operatorname{co} X \subset K$. However, $\operatorname{co} X$ is convex and contains X, so $K \subset \operatorname{co} X$. We have proved that $\operatorname{co} X = K$. $\qquad\qquad\qquad\qquad\qquad\qquad\qquad\qquad\qquad\qquad\qquad\qquad\qquad\quad\;\Box$

Corollary 7.1. *A subset X of a vector space E is convex if and only if $\operatorname{co} X = X$.*

Exercise 7.3. Let X be a convex subset of a vector space E, f a convex function defined on X and $x = \sum_{i=1}^{n} \lambda_i x_i$ a convex combination of the points $x_i \in X$. Show that

$$f(x) \le \sum_{i=1}^{n} \lambda_i f(x_i).$$

If E is a normed vector space, $\phi_1, \ldots, \phi_s \in E^*$, the dual of E, and $\alpha_1, \ldots, \alpha_s \in \mathbb{R}$, then we say that

$$P = \{x \in E : \phi_i(x) \le \alpha_i, i = 1, \ldots, s\},$$

is a *(convex) polyhedron* in E. P is clearly a closed convex subset of E; P may be empty and is not necessarily bounded. A particular example is a *closed cube* in $\mathbb{R}^n$ centred on a point a and of side-length 2ϵ:

$$\bar{C}^n(a, \epsilon) = \{x \in \mathbb{R}^n : a_i - \epsilon \le x_i \le a_i + \epsilon, i = 1, \ldots, s\}.$$

If $\phi_i = x_i$ for $i = 1, \ldots, n$, then

$$\bar{C}^n(a, \epsilon) = \{x \in \mathbb{R}^n : \phi_i(x) \le a_i + \epsilon, -\phi_i(x) \le -a_i + \epsilon, i = 1, \ldots, n\}$$

and so $\bar{C}^n(a, \epsilon)$ is a polyhedron.

(If we replace the inequalities by strict inequalities, then we obtain an open cube, i.e., if $\epsilon > 0$ and $a \in \mathbb{R}^n$, then the subset of $\mathbb{R}^n$

$$C^n(a, \epsilon) = \{x \in \mathbb{R}^n : a_i - \epsilon < x_i < a_i + \epsilon, i = 1, \ldots, s\}$$

is called an *open cube*.)

Exercise 7.4. A closed ball in $\mathbb{R}^n$ for the norm $\|\cdot\|_\infty$ is a closed cube, hence a polyhedron. Show that a closed ball for the norm $\|\cdot\|_1$ is also a polyhedron.

Let S be a convex subset of a normed vector space E and $x \in S$. If $x = \frac{x_1 + x_2}{2}$, with $x_1, x_2 \in S$, implies that $x_1 = x_2 = x$, then x is said to be an extreme point of S. In the case of a polyhedron, we usually call such a point a *vertex*. Clearly an extreme point lies on the boundary of a convex set. In the appendix to this chapter we show that, in the case of a finite-dimensional normed vector space, a bounded nonempty polyhedron has a finite number of vertices and is the convex hull of these vertices.

Exercise 7.5. We know that a closed ball for the norm $\| \cdot \|_1$ in $\mathbb{R}^n$ is a polyhedron (see Exercise 7.4). What are its vertices?

7.2 Continuity of Convex Functions

It is natural to ask whether a convex function is continuous or not. This is in general not true. For example, if we define f on $[0, 1]$ by

$$f(x) = \begin{cases} 0 & x \in [0, 1) \\ 1 & x = 1 \end{cases},$$

then f is convex, but not continuous. However, f is continuous on the interior of $[0, 1]$. We will explore this question in more depth. The infinite-dimensional and finite-dimensional cases are different. We will first consider the former. We will need to do some preliminary work.

We recall that a spanning set of a vector space V is a subset S of V such that any element $v \in V$ can be written as a linear combination of elements of S, i.e., there are elements $s_1, \ldots, s_n \in S$ and $a_1, \ldots, a_n \in \mathbb{K}$, the ground field, such that

$$v = a_1 s_1 + \cdots + a_n s_n.$$

A vector space is finite-dimensional if it contains a finite spanning set. A subset T of V is linearly independent if distinct linear combinations of elements of T produce distinct members of V. A basis of a vector space is a spanning set which is also linearly independent. It is not difficult to see that a basis is a maximal linearly independent set, and vice versa. In elementary linear algebra courses one learns that all finite-dimensional vector spaces, other than $\{0\}$, have a basis. We will now show that this (and a little more) is the case for any vector space.

Exercise 7.6. Show that a subset of a vector space is a basis if and only if it is a maximal linearly independent set.

Theorem 7.1. *Let S be a linearly independent subset of a vector space V. Then there is a basis $\mathcal{B}$ of V such that $S \subset \mathcal{B}$.*

Proof. Let $\mathcal{P}$ be the class of all linearly independent subsets of V containing S. $\mathcal{P}$ is partially ordered by inclusion. If $\mathcal{C}$ is a chain in $\mathcal{P}$, then the union of all sets of $\mathcal{C}$ is an upper bound of $\mathcal{C}$. As every chain has an upper bound, by Zorn's lemma $\mathcal{P}$ has a maximal element M. The subset M is a maximal linearly independent set (and hence a basis) containing S.   □

Corollary 7.2. *A vector space V, other than $\{0\}$, has a basis.*

Proof. It is sufficient to take $S = \{x\}$, with $x \neq 0$, in the theorem.   □

Corollary 7.3. *If A is a subspace of a vector space V, then there is a subspace B such that $A \oplus B = V$.*

Proof. If $A = V$ or $A = \{0\}$, then the result is trivial, so suppose that $A \neq V$ and $A \neq \{0\}$. From the previous corollary A has a basis $\mathcal{B}_1$. Using the theorem we see that there is a nonempty subset of independent vectors $\mathcal{B}_2$ such that $\mathcal{B} = \mathcal{B}_1 \cup \mathcal{B}_2$ is a basis of V. If we set $B = \text{Vect}(\mathcal{B}_2)$, the subspace generated by $\mathcal{B}_2$, then clearly $V = A \oplus B$.   □

Having done the preliminary work, we are now in a position to handle the question of convex functions defined on subsets of infinite-dimensional normed vector spaces.

Theorem 7.2. *If O is a nonempty open convex subset of an infinite-dimensional normed vector space E, then there is a non-continuous convex function defined on O.*

Proof. We will first show that there is a non-continuous convex function defined on E. For this, it is sufficient to show that there is a non-continuous linear form defined on E. As E is infinite-dimensional, there is an infinite sequence $(x_i)_{i=1}^{\infty}$ whose elements form an independent subset of E. Dividing by the norm if necessary, we may suppose that all the x_i are of norm 1. We define a linear form l on $A = \text{Vect}(x_1, x_2, \ldots)$ by setting $l(x_i) = 2^i$. From Corollary 7.3, there is a subspace B of E such that $E = A \oplus B$. If we set $l(y) = 0$ for $y \in B$, then we obtain an extension of l to E, which is linear but not continuous.

Now let O be any nonempty open convex subset of E. If we restrict l to O then l is convex on O. Let $a \in O$ and $r > 0$ be such that the open ball $B(a, r)$ is included in O. The sequence $(a + \frac{r}{i} x_i)_{i=1}^{\infty}$ lies in O and converges to a. However, $\lim_{i \to \infty} l(a + \frac{r}{i} x_i) = \infty$ and so l is not continuous at a. This ends the proof.   □

We will now consider the continuity of a convex function defined on an open convex subset of a finite-dimensional normed vector space E. As indicated earlier, this case is different from the infinite-dimensional case.

Lemma 7.1. *If P is a bounded nonempty polyhedron in a finite-dimensional normed vector space E and f is a convex function defined on E, then f has an upper bound on P.*

Proof. From the appendix to this chapter we know that the number of vertices of P is finite and that P is the convex hull of these vertices. If $x \in P$ and $x_1, \ldots, x_n$ are the vertices of P, then $x = \sum_{i=1}^{n} \lambda_i x_i$, with $\lambda_i \in [0, 1]$ and $\sum_{i=1}^{n} \lambda_i = 1$. We have

$$f(x) = f\left(\sum_{i=1}^{n} \lambda_i x_i\right) \leq \sum_{i=1}^{n} \lambda_i f(x_i) \leq \sum_{i=1}^{n} \lambda_i \max f(x_i) = \max f(x_i).$$

This ends the proof.   □

Theorem 7.3. *Let X be a convex subset of a finite-dimensional normed vector space E and $f : X \longrightarrow \mathbb{R}$ convex. If $x \in \mathrm{int}\, X$, then f is continuous at x.*

Proof. We will first prove the result for the case $E = \mathbb{R}^n$ with the norm $\|\cdot\| = \|\cdot\|_\infty$. Let $\bar{B}(x, r)$ be a closed ball centred on x and lying in $\mathrm{int}\, X$. Given that the norm is $\|\cdot\|_\infty$, $\bar{B}(x, r)$ is a closed cube and so a polyhedron. From the lemma f has an upper bound on $\bar{B}(x, r)$: there exists k such that $f(x) \leq k$ for all $x \in \bar{B}(x, r)$. Let $h \neq 0$ be such that $\|h\| \leq r$. Then, for $\alpha \in [0, 1]$, $x + \alpha h \in \bar{B}(x, r)$ and

$$f(x + \alpha h) = f((1 - \alpha)x + \alpha(x + h)) \leq (1 - \alpha)f(x) + \alpha f(x + h). \quad (7.1)$$

Also,

$$f(x) = f\left(\frac{1}{1 + \alpha}(x + \alpha h) + \frac{\alpha}{1 + \alpha}(x - h)\right) \leq \frac{1}{1 + \alpha}(f(x + \alpha h) + \alpha f(x - h)),$$

from which we obtain

$$(1 + \alpha)f(x) \leq f(x + \alpha h) + \alpha f(x - h). \quad (7.2)$$

Using this inequality we get

$$f(x + \alpha h) \geq (1 + \alpha)f(x) - \alpha f(x - h) \geq (1 + \alpha)f(x) - \alpha k$$

and, with $\alpha = 1$,

$$f(x + h) \geq 2f(x) - k.$$

Therefore f has a lower bound on $\bar{B}(x, r)$. As f has both an upper and a lower bound on $\bar{B}(x, r)$, f is bounded on $\bar{B}(x, r)$: we can find a constant d such that $|f(u)| \leq d$ for all $u \in \bar{B}(x, r)$.

Now, from the inequalities (7.1) and (7.2) we deduce

$$\alpha(f(x) - f(x - h)) \leq f(x + \alpha h) - f(x) \leq \alpha(f(x + h) - f(x))$$

and so

$$|f(x + \alpha h) - f(x)| \leq 2\alpha d.$$

Suppose now that $\|u\| = s \leq \frac{r}{2}$, with $s \neq 0$. If we set $h = \frac{r}{2s}u$, then $\|h\| = \frac{r}{2}$, $u = \frac{2s}{r}h$ and

$$|f(x + u) - f(x)| = \left|f\left(x + \frac{2s}{r}h\right) - f(x)\right| \leq \frac{4s}{r}d$$

and the continuity of f at x follows.

Let us now consider the general case. If E is a finite-dimensional normed vector space and $(v_i)_{i=1}^n$ a basis of E, then the mapping

$$\phi : \mathbb{R}^n \longrightarrow E, (x_1, \ldots, x_n) \longmapsto x_1 v_1 + \cdots + x_n v_n$$

is a normed vector space isomorphism for any norm of $\mathbb{R}^n$. Let f be a convex function defined on a convex subset X of E and $x \in \text{int } X$. Then $\tilde{X} = \phi^{-1}(X)$ is a convex subset of $\mathbb{R}^n$ and $\tilde{x} = \phi^{-1}(x) \in \text{int } \tilde{X}$. If we set $\tilde{f} = f \circ \phi$, then $\tilde{f}$ is a convex function defined on $\tilde{X}$ and so is continuous at $\tilde{x}$. As $f = \tilde{f} \circ \phi^{-1}$, f is continuous at x. This completes the proof. $\qquad\square$

Corollary 7.4. *If a convex function is defined on an open subset of a finite-dimensional normed vector space, then it is continuous.*

7.3 Differentiable Convex Functions

It is not always easy to use the definition of a convex function to decide whether a given function is convex or not. However, if a function is differentiable, then there are equivalent conditions which are often easier to use.

Theorem 7.4. *Let O be an open subset of a normed vector space E and f a real-valued differentiable function defined on O. If $X \subset O$ is convex and $x, y \in X$, then the following conditions are equivalent:*

(a) f *is convex on* X;
(b) $f(y) - f(x) \geq f'(x)(y - x)$;
(c) $(f'(y) - f'(x))(y - x) \geq 0$.

Proof. (a) $\Longrightarrow$ (b) Let $\lambda \in (0, 1)$. As f is convex,

$$f(x + \lambda(y - x)) \leq (1 - \lambda)f(x) + \lambda f(y).$$

If we subtract $f(x)$ from both sides of the inequality and use the differentiability of f, we get

$$\lambda f'(x)(y - x) + o(\lambda(y - x)) \leq \lambda(f(y) - f(x))$$

and, dividing by λ,

$$f'(x)(y - x) + \frac{o(\lambda(y - x))}{\lambda} \leq f(y) - f(x).$$

Letting λ go to 0, we obtain (b).
(b) $\Longrightarrow$ (c) We have

$$f(y) - f(x) \geq f'(x)(y - x) \Longrightarrow f(x) - f(y) \leq -f'(x)(y - x)$$

and

$$f(x) - f(y) \geq f'(y)(x - y) \implies f(y) - f(x) \leq f'(y)(y - x).$$

On adding the inequalities we obtain (c)

(c) $\implies$ (a) Let us fix $x, y \in X$ and set $\phi(t) = f(x + t(y - x))$. Then ϕ is defined and differentiable on an open interval I of $\mathbb{R}$ containing $[0, 1]$. For $t \in I$

$$\dot{\phi}(t) = f'(x + t(y - x))(y - x).$$

By hypothesis, if $s, t \in [0, 1]$, then

$$\Big(f'(x + t(y - x)) - f'(x + s(y - x))\Big)(x + t(y - x)) - (x + s(y - x)) \geq 0$$

i.e.,

$$\Big(f'(x + t(y - x)) - f'(x + s(y - x))\Big)(t - s)(y - x) \geq 0.$$

For $t > s$ this gives us

$$\Big(f'(x + t(y - x)) - f'(x + s(y - x))\Big)(y - x) \geq 0,$$

i.e., $\dot{\phi}(t) - \dot{\phi}(s) \geq 0$. Therefore $\dot{\phi}$ is monotone and hence Riemann integrable on any compact interval contained in $[0, 1]$. Let $\lambda \in (s, 1)$. Integrating with respect to t between λ and 1 we obtain

$$\phi(1) - \phi(\lambda) - \dot{\phi}(s)(1 - \lambda) \geq 0.$$

We now integrate this expression with respect to s between 0 and λ to obtain

$$(\phi(1) - \phi(\lambda))\lambda - (1 - \lambda)(\phi(\lambda) - \phi(0)) \geq 0,$$

or

$$\lambda\phi(1) + (1 - \lambda)\phi(0) - \phi(\lambda) \geq 0,$$

i.e.,

$$\lambda f(y) + (1 - \lambda)f(x) \geq f(\lambda y + (1 - \lambda)x).$$

Therefore f is convex. $\square$

We have a similar result for strictly convex functions, namely

Theorem 7.5. *Let O be an open subset of a normed vector space E and f a real-valued differentiable function defined on O. If $X \subset O$ is convex and $x, y \in X$ with $x \neq y$, then the following conditions are equivalent:*

(a) *f is strictly convex on X;*

(b) $f(y) - f(x) > f'(x)(y - x);$
(c) $(f'(y) - f'(x))(y - x) > 0.$

Proof. To prove (b) $\implies$ (c) and (c) $\implies$ (a) it is sufficient to slightly modify the corresponding arguments of the previous theorem. To prove that (a) $\implies$ (b) we proceed as follows. Let $\omega \in (0, 1)$ and $\theta \in (0, \omega)$. Then we can write

$$x + \theta(y - x) = \frac{\omega - \theta}{\omega} x + \frac{\theta}{\omega}(x + \omega(y - x))$$

and so

$$f(x + \theta(y - x)) \le \frac{\omega - \theta}{\omega} f(x) + \frac{\theta}{\omega} f(x + \omega(y - x)).$$

Therefore

$$f(x + \theta(y - x)) - f(x) \le \frac{\theta}{\omega}(f(x + \omega(y - x)) - f(x))$$

and

$$\frac{f(x + \theta(y - x)) - f(x)}{\theta} \le \frac{f(x + \omega(y - x)) - f(x)}{\omega}.$$

Letting θ go to 0 we obtain

$$f'(x)(y - x) \le \frac{f(x + \omega(y - x)) - f(x)}{\omega}.$$

However,

$$\frac{f(x + \omega(y - x)) - f(x)}{\omega} < \frac{1}{\omega}((1 - \omega) f(x) + \omega f(y) - f(x)) = f(y) - f(x),$$

and hence (b). $\qquad\square$

Remark. From (c) in Theorems 7.4 and 7.5 we deduce that, if f is differentiable on an open interval I of $\mathbb{R}$, then f is convex (resp. strictly convex) if and only if the derivative of f is increasing (resp. strictly increasing) on I.

Example. Let $A \in \mathcal{M}_n(\mathbb{R})$ be symmetric, $b \in \mathbb{R}^n$ and $f : \mathbb{R}^n \longrightarrow \mathbb{R}$ be defined by

$$f(x) = \frac{1}{2} x^t A x - b^t x.$$

Then

$$f(y) - f(x) - f'(x)(y - x) = \left(\frac{1}{2} y^t A y - b^t y\right) - \left(\frac{1}{2} x^t A x - b^t x\right)$$

$$-(Ax - b)^t(y - x)$$

$$= \frac{1}{2}y^t Ay + \frac{1}{2}x^t Ax - x^t Ay$$

$$= \frac{1}{2}(y - x)^t A(y - x).$$

It follows that f is convex (resp. strictly convex) if and only if the matrix A is positive (resp. positive definite).

Exercise 7.7. Let $a \in \mathbb{R}$ and $f_a : \mathbb{R}^2 \longrightarrow \mathbb{R}$ be defined by

$$f_a(x, y) = x^2 + y^2 + axy - 2x - 2y.$$

For what values of a is f_a convex? strictly convex?

We have seen that, if E is a Hilbert space, K a closed convex subset of E and $d_K(x)$ the distance from the point x to K, then the function d_K^2 is of class C^1 and

$$(d_K^2)'(x)h = 2\langle x - P_K(x), h\rangle,$$

where $P_K(x)$ is the projection of x on K. We can say more.

Theorem 7.6. *The function d_K^2 is convex.*

Proof. We have

$$((d_K^2)'(x+h) - (d_K^2)'(x))h = 2\langle x + h - P_K(x+h), h\rangle - 2\langle x - P_K(x), h\rangle$$

$$= 2\langle h - P_K(x+h) + P_K(x), h\rangle$$

$$= 2\|h\|^2 - 2\langle P_K(x+h) - P_K(x), h\rangle$$

$$\geq 2\|h\|^2 - 2\|P_K(x+h) - P_K(x)\|\|h\| \geq 0,$$

because P_K is 1-Lipschitz. It follows that d_K^2 is convex. $\square$

If a function is 2-differentiable, then we may use the second differential to determine whether the function is convex or not. This is often easier to use than the (first) differential. Let O be an open subset of a normed vector space E and f a real-valued 2-differentiable function defined on O. For $x \in O$ and $h \in E$ we set

$$Q_f(x)(h) = f^{(2)}(x)(h, h).$$

(Notice that $Q_f(x)$ is a quadratic form.)

Theorem 7.7. *Let O be an open subset of a normed vector space E, $X \subset O$ convex and $f : O \longrightarrow \mathbb{R}$ 2-differentiable. Then*

(a) *f is convex on X, if and only if $Q_f(x)$ is positive for all $x \in X$;*
(b) *f is strictly convex on X, if $Q_f(x)$ is positive definite for all $x \in X$.*

Proof. Let $x, y \in X$. From Theorem 5.3 there exists z in the segment (x, y) such that

$$f(y) - f(x) - f'(x)(y - x) = \frac{1}{2}Q_f(z)(y - x).$$

If Q_f is positive on X, then $Q_f(z)(y - x) \geq 0$ and so from Theorem 7.4 f is convex. If Q_f is positive definite on X, then $Q_f(z)(y - x) > 0$ if $y \neq x$ and so from Theorem 7.5 f is strictly convex.

Suppose now that f is convex. Let $x \in X$ and $B(x, r)$ be an open ball lying in O. If $h \in B(0, r)$ and $t \in (0, 1]$, then using Theorem 5.1 we have

$$0 \leq f(x + th) - f(x) - f'(x)(th) = \frac{1}{2}Q_f(x)(th) + o(\|th\|^2)$$

$$= t^2 \left(\frac{1}{2}Q_f(x)(h) + \frac{o(\|th\|^2)}{t^2} \right).$$

As $\lim_{t \to 0} \frac{o(\|th\|^2)}{t^2} = 0$, $Q_f(x)(h) \geq 0$. If $h' \in E \setminus B(0, r)$, then there exists $s \neq 0$ such that $sh' \in B(0, r)$ and

$$Q_f(x)(h') = \frac{1}{s^2}Q_f(x)(sh') \geq 0.$$

Hence $Q_f(x)(h) \geq 0$ for all $h \in E$.   □

Remark. A function f may be strictly convex without the quadratric form Q_f being positive definite at all points. For example, if f is the real-valued function defined on $\mathbb{R}$ by $f(x) = x^4$, then $Q_f(0) = 0$. However,

$$(x + h)^4 - x^4 = x^4 + 4x^3h + 6x^2h^2 + 4xh^3 + h^4 - x^4$$

$$= f'(x)h + h^2(6x^2 + 4xh + h^2)$$

$$= f'(x)h + h^2(2x^2 + (2x + h)^2) > f'(x)h$$

if $h \neq 0$. Therefore f is strictly convex.

Exercise 7.8. Show that the function $x \longmapsto e^{\alpha x}$, with $\alpha \in \mathbb{R}^*$, is strictly convex on $\mathbb{R}$ and that the functions $x \longmapsto x \ln x$ and $x \longmapsto -\ln x$ are strictly convex on $\mathbb{R}^*_+$.

Theorems 7.4 and 7.5 enable us to generalize the concept of convexity to differentiable functions not necessarily defined on convex sets. Let O be an open subset of a normed vector space E and f a real-valued differentiable function defined on O. If X is a subset of O (not necessarily convex) and

$$f(y) - f(x) \geq f'(x)(y - x)$$

for all $x, y \in X$, then we will say that f is generalized convex or *g-convex* on X. If

$$f(y) - f(x) > f'(x)(y - x)$$

for all $x, y \in X$ when $x \neq y$, then we will say that f is strictly generalized convex or *strictly g-convex* on X. From Theorems 7.4 and 7.5, for a differentable function defined on a convex set convexity (resp. strict convexity) and g-convexity (resp. strict g-convexity) are equivalent.

7.4   Extrema of Convex Functions

Let us first consider a convex real-valued function defined on a convex subset of a vector space (not necessarily normed). Here of course there is no notion of local extremum.

Proposition 7.3. *Let X be a convex subset of a vector space V and $f : X \longrightarrow \mathbb{R}$ a convex function. If x and y are minima of f, then any point on the segment joining x to y is also a minimum. If f is strictly convex, then f can have at most one minimum.*

Proof. If x and y are distinct minima of f and $z = \lambda x + (1 - \lambda)y \in (x, y)$, then

$$f(z) = f(\lambda x + (1 - \lambda)y) \leq \lambda f(x) + (1 - \lambda)f(y) = f(x) \leq f(z),$$

hence z is a minimum.

Suppose now that f is strictly convex and that x and y are distinct minima. If $z = \lambda x + (1 - \lambda)y$, with $\lambda \in (0, 1)$, then

$$f(z) = f(\lambda x + (1 - \lambda)y) < \lambda f(x) + (1 - \lambda)f(y) = f(x),$$

which contradicts the minimality of x. Therefore f can have at most one minimum.   □

Remark. From what we have just seen, a convex function has no minimum, one minimum or an infinite number of minima.

Example. Let $n \in \mathbb{N} \setminus \{0, 1\}$,

$$A_n = \left\{ (x_1, \ldots, x_n) \in \mathbb{R}^n : x_1 \geq 0, \ldots, x_n \geq 0, \sum_{i=1}^{n} x_i = 1 \right\}$$

and $f : R^n \longrightarrow \mathbb{R}$ be defined by

$$f(x_1, \ldots, x_n) = \sum_{i=1}^{n} x_i^2.$$

As A_n is closed and f continuous, f has a minimum on A_n. It is easy to see that A_n is a convex polyhedron. The Hessian matrix of f at any point x is $2I_n$, therefore f is strictly convex on $\mathbb{R}^n$ and so on A_n. It follows that the minimum is unique. Suppose that a is the minimum and there are coordinates a_i and a_j with $a_i \neq a_j$. If a' is the element of $\mathbb{R}^n$ obtained from a by permuting a_i and a_j, then $a' \neq a$ and $f(a') = f(a)$, which contradicts the uniqueness of the minimum. Therefore all the coordinates of a are the same and the minimum is the point $x = (\frac{1}{n}, \ldots, \frac{1}{n})$.

Exercise 7.9. Let

$$K = \{(x, y) \in \mathbb{R}^2 : x \geq 0, y \geq 0, 2x + y \leq 1, x + 2y \leq 1\}$$

and $f : \mathbb{R}^2 \longrightarrow \mathbb{R}$ be defined by

$$f(x, y) = x^2 - xy + y^2 - x - y.$$

Show that f has a unique minimum on K which lies on the boundary δK of K. Prove that, if (x, y) is the minimum, then $x = y$. Find the minimum.

Let us now consider normed vector spaces.

Proposition 7.4. *Let E be a normed vector space, X a convex subset of E and f a real-valued convex function defined on X. If $x \in X$ is a local minimum, then x is a global minimum.*

Proof. Let x be a local minimum and $y \in X$. For $\lambda \in (0, 1)$, $y_\lambda = x + \lambda (y - x) \in X$. If λ is sufficiently small, then we have

$$f(x) \leq f(y_\lambda) \leq \lambda f(y) + (1 - \lambda) f(x) \Longrightarrow \lambda f(x) \leq \lambda f(y).$$

It follows that $f(x) \leq f(y)$. $\qquad\square$

We now turn to differentiable functions.

Proposition 7.5. *Let O be an open subset of a normed vector space E, f a real-valued differentiable function defined on O and X a nonempty subset of O. If f is g-convex on X, $x \in X$ and for all $y \in X$*

$$f'(x)(y - x) \geq 0,$$

then x is a global minimum of f restricted to X. If, for $y \in X$, with $y \neq x$,

$$f'(x)(y - x) > 0,$$

then x is a unique global minimum.

Proof. As f restricted to X is g-convex, we have

$$f(y) - f(x) \geq f'(x)(y - x) \geq 0$$

for all $y \in X$, hence the first result. The second is proved in the same way. □

Remark. If x is a critical point, then $f'(x) = 0$ and so x is a global minimum.

Exercise 7.10. Consider the function $f : \mathbb{R}^2 \longrightarrow \mathbb{R}$ defined by

$$f(x, y) = 2x^2 + 3xy + 2y^2 - 2x - y$$

Show that f is strictly convex. Calculate the partial derivatives of f and so find the unique minimum of f.

Convexity is a subject which has many applications and is also extremely interesting in its own right. The book by Barvinok [2] is not only a very good introduction to the subject but also contains many advanced results. The books by Borwein and Lewis [5] and Schneider [22] are also good references.

Appendix: Convex Polyhedra

In this appendix we will look a little more closely at convex polyhedra. Our main aim is to show that a polyhedron in a finite-dimensional normed vector space is the convex hull of its vertices. Let P be a polyhedron in a normed vector space E:

$$P = \{x \in E : \phi_i(x) \leq \alpha_i, i = 1, \ldots, s\}.$$

If ϕ is a nonzero element of E^* and $\alpha \in \mathbb{R}$ is such that $P \subset \{x \in E : \phi(x) \leq \alpha\}$, then the subset of E

$$F = \{x \in P : \phi(x) = \alpha\}$$

is said to be a *face* of P. As

$$F = \{x \in P : \phi(x) \leq \alpha, -\phi(x) \leq -\alpha\},$$

F is a polyhedron. For example, in the case of the closed cube $\bar{C}^n(a, \epsilon)$ defined earlier in this chapter, the subset

$$F = \{x \in \bar{C}^n(a, \epsilon) : x_1 = a_1 + \epsilon\}$$

is a face. The boundary of a polyhedron is a subset of a union of faces:

$$\partial P \subset \cup_{i=1}^{s}\{x \in P : \phi_i(x) = \alpha_i\}.$$

The following result is elementary.

Proposition 7.6. *If P is a polyhedron in a normed vector space E and $a \in E$, then the translation $a + P$ of P is also a polyhedron. In addition, x is a vertex of P if and only if $a + x$ is a vertex of $a + P$ and F is a face of P if and only if $a + F$ is a face of $a + P$.*

We now consider the case where E is finite-dimensional. In particular, we will see that here the number of vertices of a polyhedron is finite.

Theorem 7.8. *Let E be a normed vector space, with $\dim E = n < \infty$, and $P \subset E$ a nonempty polyhedron:*

$$P = \{x \in E : \phi_i(x) \leq \alpha_i, i = 1, \ldots, s\}.$$

For $x \in P$ we note

$$A_x = \{i : \phi_i(x) = \alpha_i\} \qquad and \qquad V_x = Vect(\phi_i : i \in A_x).$$

Then x is vertex of P if and only if $V_x = E^*$.

Proof. Suppose that $V_x = E^*$ and let $x = \frac{x_1 + x_2}{2}$, with $x_1, x_2 \in P$. For $i \in A_x$

$$\phi_i(x_1) \leq \alpha_i \qquad \text{and} \qquad \phi_i(x_2) \leq \alpha_i. \tag{7.3}$$

However, for $i \in A_x$ we also have

$$\frac{\phi_i(x_1) + \phi_i(x_2)}{2} = \phi_i \left(\frac{x_1 + x_2}{2} \right) = \phi_i(x) = \alpha_i. \tag{7.4}$$

From the equations (7.3) and (7.4) we deduce that

$$\phi_i(x_1) = \phi_i(x_2) = \alpha_i \implies \phi_i(x_1 - x_2) = 0.$$

As the linear forms ϕ_i span E^*, $\phi(x_1 - x_2) = 0$ for all $\phi \in E^*$, which implies that $x_1 - x_2 = 0$, i.e., $x_1 = x_2 = x$ and so x is a vertex.

Suppose now that $V_x \neq E^*$ and let $\phi_{i_1}, \dots, \phi_{i_t}$ be a basis of V_x. Let us complete this set to a basis $(\psi_i)_{i=1}^n$ of E^*. This basis is the dual basis of a basis $(v_i)_{i=1}^n$ of E. If $y = v_{t+1}$, then $\psi_1(y) = \cdots = \psi_t(y) = 0$ and so $\phi_i(y) = 0$ for all $i \in A_x$. If $\epsilon > 0$ is sufficiently small and $x_1 = x + \epsilon y$, $x_2 = x - \epsilon y$, then $x_1, x_2 \in P$ and $x = \frac{x_1 + x_2}{2}$. Therefore x is not a vertex.   □

Corollary 7.5. *Let P be a nonempty polyhedron in a normed vector space E of dimension $n < \infty$. Then P has a finite number of vertices.*

Proof. Let

$$P = \{x \in E : \phi_i \leq \alpha_i, i = 1, \dots, s\}.$$

If $\text{Vect}\,(\phi_1, \dots, \phi_s) \neq E^*$, then P cannot have a vertex. This is the case if $s < n$. Suppose now that $\text{Vect}\,(\phi_1, \dots, \phi_s) = E^*$. If $x \in E$ is a vertex of P, then the set $S = \{\phi_i : i \in A_x\}$ contains a basis $(\phi_{i_1}, \dots, \phi_{i_n})$ of E^* and x is the unique solution of the system

$$\phi_{i_1}(u) = \alpha_{i_1}, \dots, \phi_{i_n}(u) = \alpha_{i_n}.$$

Thus the number of vertices is bounded by the number of distinct non-ordered bases which can be obtained from the set of linear forms ϕ_i defining P. This number is clearly finite (bounded by $\binom{s}{n}$).   □

Let us return to the closed cube $\bar{C}^n(a, \epsilon)$. The point x is a vertex if and only if

$$x_1 = a_1 \pm \epsilon, \dots, x_n = a_n \pm \epsilon.$$

Thus there are 2^n vertices.

Corollary 7.6. *If x is a vertex of the polyhedron P, then $\{x\}$ is a face of P.*

Proof. Let

$$P = \{u \in E : \phi_i(u) \leq \alpha_i, i = 1, \dots, s\}$$

and x be the unique solution of the system

$$\phi_{i_1}(u) = \alpha_{i_1}, \ldots, \phi_{i_n}(u) = \alpha_{i_n}.$$

If we set

$$\phi = \sum_{j=1}^{n} \phi_{i_j} \qquad \text{and} \qquad \alpha = \sum_{j=1}^{n} \alpha_{i_j},$$

then

$$P \subset \{u \in E : \phi(u) \le \alpha\} \qquad \text{and} \qquad \{x\} = \{u \in P : \phi(u) = \alpha\}.$$

Hence $\{x\}$ is a face of P.   □

The next result is elementary, but very useful.

Proposition 7.7. *If F is a face of P and x a vertex of F, then x is a vertex of P.*

Proof. Let

$$F = \{x \in P : \phi(x) = \alpha\}$$

be a face of P. If $x = \frac{x_1 + x_2}{2}$ with $x_1, x_2 \in P$, then

$$\alpha = \phi(x) = \frac{1}{2}(\phi(x_1) + \phi(x_2)).$$

As $x_1, x_2 \in P$, we also have $\phi(x_1) \le \alpha$ and $\phi(x_2) \le \alpha$ and so

$$\phi(x_1) = \phi(x_2) = \alpha \implies x_1, x_2 \in F.$$

As x is a vertex of F, we have $x_1 = x_2 = x$ and so x is vertex of P.   □

We have shown that (at least in a finite-dimensional space) a polyhedron can have only a finite number of vertices. We have also seen that a polyhedron may not have a vertex. We will now prove a sufficient condition for a polyhedron to have a vertex.

Theorem 7.9. *If P is a bounded nonempty polyhedron in a finite-dimensional normed vector space E, then P has a vertex.*

Proof. We will prove this result by induction on $n = \dim E$. If $n = 0$, then there is nothing to prove. Suppose now that the result is true for n and consider the case $n+1$. As P is closed and bounded, P has a boundary point $\bar{x}$ (see Exercise 1.18). As the boundary is included in a union of faces, there is a face F such that $\bar{x} \in F$. Let

$$F = \{x \in P : \phi(x) = \alpha\}.$$

Then F is contained in the hyperplane

$$H = \{x \in E : \phi(x) = \alpha\}.$$

The kernel V of ϕ is a subspace of E of dimension n and $-\bar{x} + F$ is a bounded nonempty polyhedron contained in V. From the induction hypothesis, $-\bar{x} + F$ contains a vertex $\bar{w}$ and so F contains a vertex $w (= \bar{w} + \bar{x})$. From Proposition 7.7 w is a vertex of P. This finishes the induction step.   □

To close this appendix we prove a fundamental result due to Minkowski.

Theorem 7.10. *If P is a bounded nonempty polyhedron in a finite-dimensional normed vector space, then P is the convex hull of its vertices.*

Proof. Let us note S the convex hull of the vertices of P. From the previous theorem we know that $S \neq \emptyset$. As P is convex, $S \subset P$. We need to show that $P \subset S$. We will prove the result by induction on $n = \dim E$. If $n = 0$, then there is nothing to prove. Suppose now that the result is true for n and consider the case $n + 1$. Let y be an element of P.

First suppose that y belongs to a face F of P. If

$$F = \{x \in P : \phi(x) = \alpha\},$$

then F is contained in the hyperplane

$$H = \{x \in E : \phi(x) = \alpha\}.$$

The kernel V of ϕ is a subspace of E of dimension n and $-y + F$ is a bounded nonempty polyhedron contained in V. From the induction hypothesis 0 is a convex combination of the vertices of $-y + F$ and so y is a convex combination of the vertices of F. From Proposition 7.7, $y \in S$.

Now let us suppose that y is not in a face of P and let l be a line through y whose direction vector is $v \neq 0$:

$$l = \{x \in E : x = y + \lambda v, \lambda \in \mathbb{R}\}.$$

We set

$$\bar{\lambda}_1 = \sup\{\lambda' \geq 0 : y + \lambda v \in P, \lambda \leq \lambda'\} \text{ and } \bar{\lambda}_2 = \inf\{\lambda' \leq 0 : y + \lambda v \in P, \lambda \geq \lambda'\}.$$

If $\bar{x}_1 = y + \bar{\lambda}_1 v$ and $\bar{x}_2 = y + \bar{\lambda}_2 v$, then $\bar{x}_1, \bar{x}_2 \in P$, because P is closed. In addition, $\bar{x}_1, \bar{x}_2 \in \partial P$ and $\bar{x}_1 \neq \bar{x}_2$. As ∂P is contained in a union of faces, $\bar{x}_1$ and $\bar{x}_2$ belong to faces of P and so, from the induction hypothesis, are convex combinations of vertices of P. However, y lies in the segment joining $\bar{x}_1$ to $\bar{x}_2$ and so $y \in S$. This finishes the induction step and so the proof.   □

Remark. We have shown that a bounded nonempty convex polyhedron is a *convex polytope*, i.e., the convex hull of a finite number of points. In fact, the converse is also true, namely a convex polytope is a bounded convex polyhedron. A proof of this may be found in [2]. An extensive (and readable) study of convex polytopes is made in [12].

Theorem 7.10 can be generalized to infinite-dimensional spaces. The Krein–Milman theorem states that a convex compact subset of a normed vector space is the closure of the convex hull of its extreme points. A proof of this result may be found in [20].

Chapter 8
The Inverse and Implicit Mapping Theorems

In this chapter we will prove the inverse and implicit mapping theorems, which have far-reaching applications. We will begin with the inverse mapping theorem and then derive the implicit mapping theorem from it.

8.1 The Inverse Mapping Theorem

Suppose that E and F are normed vector spaces and that O and U are open subsets of E and F respectively. We recall that $f : O \longrightarrow U$ is a diffeomorphism if f is bijective and both f and f^{-1} are differentiable. Also, we say that f is a C^k-diffeomorphism if both f and f^{-1} are C^k-mappings. In Chap. 4 Sect. 4.9 we showed that, if f is a diffeomorphism, then at any point x in its domain $f'(x)$ is invertible. In addition, we proved that for f to be a C^k-diffeomorphism it is sufficient that f be of class C^k. Here we will look at what we may call an inverse question, namely what can we say about a mapping f if f' is invertible at a point a in the domain of f. Here is a first result.

Proposition 8.1. *Let E and F be Banach spaces, $O \subset E$ and $U \subset F$ open sets and $f : O \longrightarrow U$ a differentiable homeomorphism. If $a \in O$ and $f'(a)$ is invertible, then f^{-1} is differentiable at $b = f(a)$.*
If in addition f is of class C^1, then there is an open neighbourhood O' of a such that $f_{|o'}$ is a C^1-diffeomorphism onto its image.

Proof. To simplify the notation let us set $g = f^{-1}$ and $L = f'(a)$. In the first appendix to this chapter we prove that the inverse of a continuous linear bijection from one Banach space onto another is also continuous. Thus L^{-1} is continuous. If k is small, then $f(a) + k \in U$. As f is bijective, there is a unique $h \in U$ such that $f(a + h) = f(a) + k$. First we notice that

$$g(b+k)-g(b) = f^{-1}\Big(f(a)+(f(a+h)-f(a))\Big)-f^{-1}(f(a)) = a+h-a = h.$$

R. Coleman, *Calculus on Normed Vector Spaces*, Universitext,
DOI 10.1007/978-1-4614-3894-6_8, © Springer Science+Business Media New York 2012

As g is continuous, we have $\lim_{k \to 0} h = 0$. Now,

$$k = f(a + h) - f(a) = L(h) + \|h\|\epsilon(h),$$

where $\lim_{h \to 0} \epsilon(h) = 0$. From this we obtain

$$L^{-1}(k) = h + \|h\|L^{-1}(\epsilon(h)) = g(b + k) - g(b) + \|h\|L^{-1}(\epsilon(h)).$$

Let us consider the term $\|h\|L^{-1}(\epsilon(h))$. We will show that

$$\|h\|L^{-1}(\epsilon(h)) = o(k).$$

As $L^{-1}(k) = h + \|h\|L^{-1}(\epsilon(h))$, we have

$$\|h\| \leq \|h\|\|L^{-1}(\epsilon(h))\| + \|L^{-1}(k)\| \leq \|h\|\|L^{-1}(\epsilon(h))\| + |L^{-1}|\|k\|.$$

If k is small, then so is h, which implies that $L^{-1}(\epsilon(h))$ is small. We can write

$$\|h\| \leq \frac{|L^{-1}|}{1 - \|L^{-1}(\epsilon(h))\|}\|k\| \leq M\|k\|,$$

where M is a strictly positive constant. Hence

$$\left\| \frac{\|h\|L^{-1}(\epsilon(h))}{\|k\|} \right\| \leq M\|L^{-1}(\epsilon(h))\|.$$

As the expression on the right-hand side converges to 0 when k goes to 0, we have $\|h\|L^{-1}(\epsilon(h)) = o(k)$. This gives us the first result.

Now let us turn to the second part of the proposition. Suppose that the mapping f is also of class C^1. $\mathcal{I}(E, F)$, the collection of invertible mappings in $\mathcal{L}(E, F)$, is an open subset. If U' is an open neighbourhood of $f'(a)$ in $\mathcal{I}(E, F)$ and $O' = (f')^{-1}(U')$, then O' is an open subset of E such that $f'(x)$ is invertible if $x \in O'$. If we now apply the first part of the proposition to every point $x \in O'$, then we see that f^{-1} is differentiable at every point $y \in f(O')$. Hence f restricted to O' is a diffeomorphism. As f is of class C^1, f is a C^1-diffeomorphism. $\qquad\square$

We will now weaken the hypotheses of the previous proposition. In fact, to obtain the second part of the proposition we do not need to suppose that the mapping f is a homeomorphism. We will show that f can be restricted to an open neighbourhood O' of the point a such that $f_{|_{O'}}$ is a homeomorphism onto its image.

Theorem 8.1. *(Inverse mapping theorem) Let E and F be Banach spaces, $O \subset E$ and $f : O \longrightarrow F$ of class C^1. If $a \in O$ and $f'(a)$ is invertible, then there is an open neighbourhood O' of a such that $f_{|_{O'}}$ is a C^1-diffeomorphism onto its image.*

Proof. Let us first suppose that $E = F$, $f(a) = a = 0$ and $f'(a) = \mathrm{id}_E$. As f is of class C^1, there is a closed ball $\bar{B}(0, r)$, with $r > 0$, included in O such that for all $x \in \bar{B}(0, r)$

$$|f'(x) - \mathrm{id}_E| \le \frac{1}{2}.$$

If we set $g(x) = f(x) - x$, then $g'(x) = f'(x) - \mathrm{id}_E$. Using Corollary 3.2 we obtain

$$\|f(x) - x\| = \|g(x) - g(0)\| \le \frac{1}{2}\|x\|.$$

Now let us choose an element $y \in B(0, \frac{r}{2})$. Then

$$\|y + x - f(x)\| \le \|y\| + \|x - f(x)\| \le \|y\| + \frac{1}{2}\|x\| < r$$

and so $y + x - f(x) \in \bar{B}(0, r)$.

We now define the mapping $\phi_y : \bar{B}(0, r) \longrightarrow \bar{B}(0, r)$ by

$$\phi_y(x) = y + x - f(x).$$

Then, using Corollary 3.2 again, for $w, z \in \bar{B}(0, r)$ we have

$$\|\phi_y(w) - \phi_y(z)\| = \|w - f(w) - z + f(z)\| = \|g(w) - g(z)\| \le \frac{1}{2}\|w - z\|$$

and so ϕ_y is a contraction. From the contraction mapping theorem (Appendix 2), ϕ_y has a unique fixed point $x \in \bar{B}(0, r)$: $x = \phi_y(x) = y + x - f(x)$. As the image of ϕ_y lies in $B(0, r)$, $x \in B(0, r)$. We have shown that there is a unique $x \in B(0, r)$ such that $f(x) = y$.

If we set $\tilde{O} = f^{-1}(B(0, \frac{r}{2})) \cap B(0, r)$, then $\tilde{O}$ is an open neighbourhood of 0, $f(\tilde{O}) = B(0, \frac{r}{2})$ and $f_{|\tilde{O}}$ is a continuous bijection onto its image. In addition,

$$\|w - z\| - \|f(w) - f(z)\| \le \|w - z - f(w) + f(z)\| = \|g(w) - g(z)\| \le \frac{1}{2}\|w - z\|.$$

This implies that

$$\|w - z\| \le 2\|f(w) - f(z)\|,$$

i.e., $f_{|\tilde{O}}^{-1}$ is a 2-Lipschitz mapping and so continuous. We have shown that $f : \tilde{O} \longrightarrow B(0, \frac{r}{2})$ is a homeomorphism. Also, f is of class C^1. Applying Proposition 8.1 we obtain the result.

Now let us relax the constraints $E = F$, $f(a) = a = 0$ and $f'(a) = \mathrm{id}_E$. The mapping ψ defined on an open neighbourhood of $0 \in E$ by

$$\psi(x) = f'(a)^{-1}(f(a + x) - f(a))$$

has its image in E, and is of class C^1; also $\psi(0) = 0$ and $\psi'(0) = \mathrm{id}_E$. From what we have just seen, there are neighbourhoods O and U of the origin such that $\psi : O \longrightarrow U$ is a C^1-diffeomorphism. For the restriction of f to $O' = a + O$ we have

$$f_{|_{O'}}(x) = f(a) + f'(a)\psi(x - a)$$

and $f_{|_{O'}}$ is a bijection onto $V = f(a) + f'(a)U$. As the inverse of $f_{|_{O'}}$ may be written

$$f_{|_{O'}}^{-1}(y) = a + \psi^{-1}\left(f'(a)^{-1}(y - f(a))\right),$$

$f_{|_{O'}}^{-1}$ is of class C^1. Hence $f_{|_{O'}}$ is a C^1-diffeomorphism.   □

Remarks. 1. Under the conditions of the theorem, $f_{|_{O'}}$ is a C^1-diffeomorphism onto its image. In fact, if f is of class C^k, then $f_{|_{O'}}$ is a C^k-diffeomorphism.

2. If a mapping f is such that each point in its domain has an open neighbourhood O such that f restricted to O defines a diffeomorphism onto its image, then we say that f is a *local diffeomorphism*.

Example. Consider the mapping

$$f : \mathbb{R}^2 \setminus \{(0,0)\} \longrightarrow \mathbb{R}^2, (r, \theta) \longmapsto (r\cos\theta, r\sin\theta).$$

Then

$$J_f(r, \theta) = \begin{pmatrix} \cos\theta & -r\sin\theta \\ \sin\theta & r\cos\theta \end{pmatrix}$$

and $\det J_f(r, \theta) = r \neq 0$. It follows that $f'(r, \theta)$ is invertible for all $(r, \theta) \in \mathbb{R}^2 \setminus \{(0,0)\}$. The continuity of the entries in the Jacobian matrix imply that f is a C^1-mapping. Hence f is a local diffeomorphism; however, f is not bijective and so not a diffeomorphism.

The next example is a little more complex.

Example. Let E be a finite-dimensional normed vector space and f a C^1-mapping from E into itself. In addition, we suppose that

$$\|x - y\| \leq \|f(x) - f(y)\|. \tag{8.1}$$

We will show that f is a C^1-diffeomorphism. Clearly f is injective. If $y \in \overline{f(E)}$, then there is a sequence $(x_n) \in E$ such that $\lim f(x_n) = y$. As the sequence $(f(x_n)) \in E$ is a Cauchy sequence, so is the sequence (x_n). However, E is finite-dimensional and thus a Banach space and so the sequence (x_n) has a limit x. The continuity of f implies that $f(x) = y$ and it follows that $f(E)$ is closed. We will now use the inverse mapping theorem to show that $f(E)$ is also open. Let $x \in E$. As f is differentiable, we can write

$$f(x + h) - f(x) = f'(x)h + \|h\|\epsilon(h),$$

where $\lim_{h \to 0} \epsilon(h) = 0$. Using the inequality (8.1) we obtain

$$\|h\| \le \|f'(x)h\| + \|h\|\|\epsilon(h)\|.$$

There exists an $\eta > 0$ for which $\|\epsilon(h)\| \le \frac{1}{2}$ when $\|h\| \le \eta$. In this case

$$\|f'(x)h\| \ge \frac{1}{2}\|h\|.$$

Let $w \in E$ with $w \ne 0$. Then

$$\|f'(x)\eta\frac{w}{\|w\|}\| \ge \frac{\eta}{2} \implies \|f'(x)w\| \ge \frac{\|w\|}{2} \implies f'(x)w \ne 0.$$

Therefore $f'(x)$ is injective and so invertible. From the inverse mapping theorem $f(x)$ has an open neighbourhood lying in $f(E)$. It follows that $f(E)$ is an open set. As E is connected (Proposition 1.10) and $f(E)$ is a nonempty subset of E which is open and closed, we have $f(E) = E$. Hence f is a bijection. It now follows that f is a C^1-diffeomorphism from E onto itself.

8.2 The Implicit Mapping Theorem

If E is normed vector space, f a real-valued function defined on a subset S of E and $c \in \mathbb{R}$, then we say that the subset $L_c = \{z \in S : f(z) = c\}$ is the *level set* of height c of f. Consider the function $f : \mathbb{R}^2 \longrightarrow \mathbb{R}, (x, y) \longmapsto x^2 + y^2$. If $c < 0$, then $L_c = \emptyset$ and if $c = 0$, then L_c contains the unique point $(0, 0)$. If $c > 0$, then L_c contains an infinite number of points. In the latter case it is natural to ask whether L_c is the graph of some function defined on a subset of $\mathbb{R}$. Suppose that we can write

$$L_c = \{(x, \phi(x)) : x \in S\}.$$

If $(x, y) \in L_c$ with $y \ne 0$, then $(x, -y) \in L_c$. This means that $y = -\phi(x)$. It follows that $y = 0$, a contradiction. In the same way we see that we cannot write

$$L_c = \{(\psi(y), y) : y \in T\}$$

for some function ψ. Therefore L_c is not the graph of a function. Suppose now that $(a, b) \in L_c$. It is easy to see that we can restrict f to an open disc D containing (a, b) such that

$$L_c \cap D = \{(x, \phi(x)) : x \in I\} \quad \text{or} \quad L_c \cap D = \{(\psi(y), y) : y \in J\},$$

where I and J are intervals of $\mathbb{R}$, $\phi(x) = \sqrt{c - x^2}$ or $\phi(x) = -\sqrt{c - x^2}$ and $\psi(x) = \sqrt{c - y^2}$ or $\psi(x) = -\sqrt{c - y^2}$. That is, even if we cannot write L_c as the graph of a function, we can do so locally. We aim to generalize this idea.

Theorem 8.2. *(Implicit mapping theorem) Let E_1, E_2 and F be Banach spaces, O an open subset of $E_1 \times E_2$ and $f : O \longrightarrow F$ a C^1-mapping. Suppose that $c \in F$ and that the set S of pairs $(x, y) \in O$ satisfying the relation*

$$f(x, y) = c$$

is not empty. If $(a, b) \in S$ and the partial differential $\partial_2 f(a, b) : E_2 \longrightarrow F$ is invertible, then there is an open neighbourhood O' of (a, b) included in O, a neighbourhood U of a in E_1 and a C^1-mapping $\phi : U \longrightarrow E_2$ such that the following statements are equivalent:

1. $(x, y) \in O'$ and $f(x, y) = c$;
2. $x \in U$ and $y = \phi(x)$.

Proof. To begin with, let us define $g : O \longrightarrow E_1 \times F$ by

$$g(x, y) = (x, f(x, y)).$$

The mapping g is of class C^1, with

$$g'(x, y)(u, v) = (u, f'(x, y)(u, v)) = (u, \partial_1 f(x, y)u + \partial_2 f(x, y)v).$$

In particular, at (a, b) we have

$$g'(a, b)(u, v) = (u, f'(a, b)(u, v)) = (u, \partial_1 f(a, b)u + \partial_2 f(a, b)v).$$

The differential $g'(a, b)$ is invertible, with

$$g'(a, b)^{-1}(u, w) = (u, \partial_2 f(a, b)^{-1}(w - \partial_1 f(a, b)u)).$$

Applying the inverse mapping theorem we obtain a neighbourhood O' of (a, b), with $O' \subset O$, such that $g_{|O'}$ is a C^1-diffeomorphism onto its image W. If $h = g_{|O'}^{-1}$, then for $(x, y) \in O'$ we have

$$(x, y) = h \circ g(x, y) = h(x, f(x, y)) = (x, h_2(x, z)),$$

where $z = f(x, y)$. The mapping h_2 is the second coordinate of the C^1-mapping $h : W \longrightarrow E_1 \times E_2$ and so is of class C^1 (Proposition 4.10). We now set

$$U = \{x \in E_1 : (x, c) \in W\}.$$

U is the inverse image of W under the inclusion mapping $i_c : E_1 \longrightarrow E_1 \times F, x \longmapsto (x, c)$ and so is open. Clearly $a \in U$. The mapping $\phi = h_2 \circ i_c$ from U into E_2, being a composition of C^1-mappings, is of class C^1 and satisfies the equivalent statements of the theorem.   □

Remark. Under the conditions of the theorem, if the equation

$$f(x, y) = c$$

has a solution, then it has an infinite number of solutions.

In general, the mapping ϕ cannot be explicitly determined. However, it is possible to find its differential at a. We have the relation $f(x, \phi(x)) = c$ for points in a neighbourhood of a, from which we obtain

$$\partial_1 f(a, b) + \partial_2 f(a, b) \circ \phi'(a) = 0.$$

This implies that

$$\phi'(a) = -\partial_2 f(a, b)^{-1} \circ \partial_1 f(a, b).$$

In fact, we can do a little better. From Proposition 3.2 we know that $\partial_2 f$ is a continuous mapping from O into $\mathcal{L}(E_2, F)$. The invertible elements of $\mathcal{L}(E_2, F)$ form an open subset $\mathcal{I}(E_2, F)$ and so there is an open subset O'' of O, containing (a, b), such that $\partial_2 f(O'') \subset \mathcal{I}(E_2, F)$. We can suppose that $O'' \subset O'$ (if not, we can take the intersection of O' and O''). For $(x, y) \in O''$

$$\partial_1 f(x, y) + \partial_2 f(x, y) \circ \phi'(x) = 0,$$

which implies that

$$\phi'(x) = -\partial_2 f(x, y)^{-1} \circ \partial_1 f(x, y).$$

Remark. If the mapping f is if class C^k, then the mapping ϕ is also of class C^k. To see why this is so, let us return to the proof of the implicit mapping theorem. Because f is of class C^k, g is also of class C^k. It follows that $h = g_{|O'}^{-1}$ is of class C^k. As h_2 is a coordinate mapping of h, h_2 is also of class C^k. To finish, ϕ is the composition of two C^k-mappings and hence of class C^k.

Let $f_1, \ldots, f_p$ be real-valued functions of class C^1 defined on some open subset O of $\mathbb{R}^{n+p}$ and consider the system of equations (S):

$$f_1(x_1, \ldots, x_n, y_1, \ldots, y_p) = c_1$$

$$\vdots$$

$$f_p(x_1, \ldots, x_n, y_1, \ldots, y_p) = c_p.$$

We can write this system as

$$f(x, y) = c. \tag{8.2}$$

From the implicit mapping theorem, if $f(a, b) = c$ and the partial differential $\partial_2 f(a, b)$ is a linear isomorphism, then in some neighbourhood of (a, b), the points (x, y) satisfying (8.2) form the graph of a C^1-mapping ϕ, defined on an open subset of $\mathbb{R}^n$ with image in $\mathbb{R}^p$, i.e., we can write

$$\phi_1(x_1, \ldots, x_n) = y_1$$
$$\vdots$$
$$\phi_p(x_1, \ldots, x_n) = y_p,$$

where the mappings are all of class C^1. To see whether $\partial_2(a, b)$ is a linear isomorphism, it is sufficient to determine whether the determinant of the matrix

$$M(x, y) = \left(\frac{\partial f_i}{\partial y_j}(x, y) \right)_{1 \le i, j \le p}$$

at (a, b) is not 0. We also have an expression for the Jacobian matrix of ϕ. If

$$N(x, y) = \left(\frac{\partial f_i}{\partial x_j}(x, y) \right)_{1 \le i \le p, 1 \le j \le n},$$

then

$$J_\phi(x) = -M(x, y)^{-1} N(x, y).$$

Example. Consider the system

$$f_1(x, y_1, y_2) = x^2 + y_1^2 + y_2^2 = 3$$
$$f_2(x, y_1, y_2) = x^2 + 3xy_1 - 2y_1 = 0.$$

$(1, -1, 1)$ is a solution of the system. In addition,

$$M(x, y_1, y_2) = \left(\frac{\partial f_i}{\partial y_j}(x, y_1, y_2) \right)_{1 \le i, j \le 2} = \begin{pmatrix} 2y_1 & 2y_2 \\ 3x - 2 & 0 \end{pmatrix}$$

and $\det M(x, y_1, y_2) = 2y_2(2 - 3x)$. As $\det M(1, -1, 1) = -2 \neq 0$, there exist mappings ϕ_1 and ϕ_2, defined on a neighbourhood of 1, such that

$$\phi_1(x) = y_1 \qquad \text{and} \qquad \phi_2(x) = y_2,$$

for solutions (x, y_1, y_2) of the system close to $(1, -1, 1)$. Because the functions of the system are of class C^∞, so are the mappings ϕ_1 and ϕ_2. We can calculate $\dot{\phi}_1(1)$ and $\dot{\phi}_2(1)$ from the expressions for $M(x, y_1, y_2)$ and $N(x, y_1, y_2)$ at $(1, -1, 1)$. We find $\dot{\phi}_1(1) = 1$ and $\dot{\phi}_2(1) = 0$.

Here is another example.

Example. Consider the equation

$$f(x_1, x_2, y) = (x_1^2 + x_2^2 + y^2) \ln(x_1 + x_2 + y) - e^{x_1 + x_2} + 1 = 0.$$

$(0, 0, 1)$ is a solution of this equation and $\frac{\partial f}{\partial y}(0, 0, 1) = 1 \neq 0$. From the implicit mapping theorem there exists a real-valued function ϕ of class C^∞, defined on a neighbourhood of $(0, 0)$, such that

$$\phi(x_1, x_2) = y$$

for solutions (x_1, x_2, y) of the equation close to $(0, 0, 1)$. The Jacobian matrix of ϕ at $(0, 0)$ is $(0\ 0)$.

Exercise 8.1. Consider the equation

$$\arctan(xy) - e^{x+y} + 1 = 0.$$

Show that $(0, 0)$ is a solution of the equation and that in a neighbourhood of $(0, 0)$ the points (x, y) satisfying the equation are the graph of a real-valued function $\phi(x)$ of class C^∞. Determine the derivative of the function ϕ at 0.

Exercise 8.2. Consider the system (E_s) of equations

$$2x + y + s(1 - x^2) = 0$$
$$-5x - 2y + s(1 - y^2) = 0.$$

Show that there are real-valued functions ϕ and ψ of class C^∞, defined on a neighbourhood of 0, such that $(\phi(s), \psi(s))$ is a solution of the system (E_s). Determine the derivatives of the functions ϕ and ψ at 0.

Exercise 8.3. Consider the system of equations (S)

$$x^2 - y^2 + wz = 0$$
$$xy - w^2 + z^2 = 0.$$

Show that $(0, 1, 1, 1)$ is a solution of (S) and that there is a neighbourhood O of $(0, 1)$ and C^∞-mappings $\phi_1 : O \longrightarrow \mathbb{R}^2$ and $\phi_2 : O \longrightarrow \mathbb{R}^2$ such that $(x, y, \phi_1(x, y), \phi_2(x, y))$ is a solution of (S) for all $(x, y) \in O$. Find the partial derivatives of ϕ_1 and ϕ_2 at $(0, 1)$.

It is interesting to consider a system of equations (S) composed of affine functions:

$$a_{11}x_1 + \cdots + a_{1n}x_n + b_{11}y_1 + \cdots + b_{1p}y_p = c_1$$

$$\vdots$$

$$a_{p1}x_1 + \cdots + a_{pn}x_n + b_{p1}y_1 + \cdots + b_{pp}y_p = c_p.$$

We may write (S) in matrix form as follows:

$$AX + BY = c$$

If $\det B \neq 0$, then B is invertible. For any c the system has a solution, for example $(0, B^{-1}c)$. The solutions are of the form $(X, \phi(X))$, with $X \in \mathbb{R}^n$ and

$$\phi(X) = -B^{-1}AX + B^{-1}c.$$

Exercise 8.4. Prove that the solutions of the system of affine functions are of the form just given.

8.3   The Rank Theorem

In this section we will only be concerned with finite-dimensional normed vector spaces. We will extensively use the inverse mapping theorem. If E and F are normed vector spaces, O an open subset of E and f a mapping from E into F which is differentiable at $a \in O$, then we call the rank of the differential $f'(a)$ the *rank of f at a* . As rk $f'(a) \leq \min(\dim E, \dim F)$, the rank of f at a cannot be greater than the minimum of the dimensions of E and F. We will be particularly interested in the case where f is differentiable at all points $x \in O$ and the rank is constant; then we will speak of the rank of f. Suppose that the rank of f is equal to $\min(\dim E, \dim F)$. If $\min(\dim E, \dim F) = \dim F$, then $f'(x)$ is surjective for all x and we say that f is a *submersion*; if $\min(\dim E, \dim F) = \dim E$, then $f'(x)$ is injective for all x and we say that f is an *immersion*.

Our aim is to look at the local structure of C^1-mappings between finite-dimensional normed vector spaces. We will first consider linear mappings, and then generalize the result.

Theorem 8.3. *Let $f : \mathbb{R}^m \longrightarrow \mathbb{R}^n$ be a linear mapping of rank r. Then there are linear isomorphisms $\alpha : \mathbb{R}^m \longrightarrow \mathbb{R}^m$ and $\beta : \mathbb{R}^n \longrightarrow \mathbb{R}^n$ such that*

$$\beta \circ f \circ \alpha(x_1, \ldots, x_m) = (x_1, \ldots, x_r, 0, \ldots, 0).$$

Proof. Let V be a complementary subspace of $\operatorname{Ker} f$, the kernel of $f: \mathbb{R}^m = V \oplus \operatorname{Ker} f$. The dimension of V is r and f restricted to V is injective. Let $u_1, \ldots, u_r$ be a basis of V and $u_{r+1}, \ldots, u_m$ a basis of $\operatorname{Ker} f$. Then $\mathcal{U} = (u_i)_{i=1}^m$ is a basis of $\mathbb{R}^m$. Now let W be a complementary subspace of $\operatorname{Im} f$, the image of f in $\mathbb{R}^n$: $\mathbb{R}^n = \operatorname{Im} f \oplus W$. The dimension of $\operatorname{Im} f$ is r and $v_1 = f(u_1), \ldots, v_r = f(u_r)$ is a basis of $\operatorname{Im} f$. If $(v_{r+1}, \ldots, v_n)$ is a basis of W, then $\mathcal{V} = (v_j)_{j=1}^n$ is a basis of $\mathbb{R}^n$. Supposing that M is the matrix of f in the bases (u_i) and (v_j), then

$$M = \begin{pmatrix} I_r & 0_{r,m-r} \\ 0_{n-r,r} & 0_{n-r,m-r} \end{pmatrix}.$$

Let $\mathcal{B}_m$ (resp. $\mathcal{B}_n$) be the standard basis of $\mathbb{R}^m$ (resp. $\mathbb{R}^n$). With A the matrix of f in the standard bases we have the relation

$$M = \mathcal{P}_{\mathcal{V}\mathcal{B}_n} A \mathcal{P}_{\mathcal{B}_m\mathcal{U}},$$

where $\mathcal{P}_{\mathcal{A}\mathcal{B}}$ is the matrix representing the basis $\mathcal{B}$ in the basis $\mathcal{A}$. If, for $x \in \mathbb{R}^m$ and $y \in \mathbb{R}^n$, we set

$$\alpha(x) = \mathcal{P}_{\mathcal{B}_m\mathcal{U}}x \qquad \text{and} \qquad \beta(y) = \mathcal{P}_{\mathcal{V}\mathcal{B}_n}y,$$

then α and β are linear isomorphisms and

$$\beta \circ f \circ \alpha(x_1, \ldots, x_m) = (x_1, \ldots, x_r, 0, \ldots, 0).$$

This completes the proof. $\qquad\qquad\qquad\qquad\qquad\qquad\qquad\qquad\qquad\qquad\qquad\qquad\qquad$ $\square$

Corollary 8.1. *Let E and F be vector spaces, with $\dim E = m$ and $\dim F = n$, and $f : E \longrightarrow F$ a linear mapping of rank r. Then there are linear isomorphisms $\bar{\alpha} : \mathbb{R}^m \longrightarrow E$ and $\bar{\beta} : F \longrightarrow \mathbb{R}^n$ such that*

$$\bar{\beta} \circ f \circ \bar{\alpha}(x_1, \ldots, x_m) = (x_1, \ldots, x_r, 0, \ldots, 0).$$

Proof. Let $\tilde{\alpha} : \mathbb{R}^m \longrightarrow E$ and $\tilde{\beta} : F \longrightarrow \mathbb{R}^n$ be linear isomorphisms. If we set $\tilde{f} = \tilde{\beta} \circ f \circ \tilde{\alpha}$, then $\tilde{f}$ is a linear mapping from $\mathbb{R}^m$ into $\mathbb{R}^n$ and $\operatorname{rk} \tilde{f} = r$. From the theorem there are linear isomorphisms $\alpha : \mathbb{R}^m \longrightarrow \mathbb{R}^m$ and $\beta : \mathbb{R}^n \longrightarrow \mathbb{R}^n$ such that

$$\beta \circ \tilde{f} \circ \alpha(x_1, \ldots, x_m) = (x_1, \ldots, x_r, 0, \ldots, 0),$$

i.e.,

$$\beta \circ \tilde{\beta} \circ f \circ \tilde{\alpha} \circ \alpha(x_1, \ldots, x_m) = (x_1, \ldots, x_r, 0, \ldots, 0).$$

Setting $\bar{\alpha} = \tilde{\alpha} \circ \alpha$ and $\bar{\beta} = \beta \circ \tilde{\beta}$, we obtain the result. $\qquad\qquad\qquad$ $\square$

Notice that, if E and F are normed vector spaces, then $\bar{\alpha}$ and $\bar{\beta}$ are diffeomorphisms and f a C^1-mapping of constant rank. We will now generalize this result to other C^1-mappings of constant rank. We will first consider submersions and immersions between euclidean spaces.

Proposition 8.2. *Let O be an open subset of $\mathbb{R}^m$, $f : O \longrightarrow \mathbb{R}^n$ a C^1-submersion and $a \in O$. There exist an open neighbourhood U of a, an open cube U' in $\mathbb{R}^m$, centred on the origin, and C^1-diffeomorphisms $\alpha : U' \longrightarrow U$ and $\beta : \mathbb{R}^n \longrightarrow \mathbb{R}^n$ such that*

$$\beta \circ f \circ \alpha(x_1, \ldots, x_n, \ldots, x_m) = (x_1, \ldots, x_n)$$

for all $x \in U'$.

Proof. The Jacobian matrix $J_f(a)$ has n independent columns; without loss of generality, let us suppose that the first n columns are independent. We set

$$y = (y_1, \ldots, y_n, y_{n+1}, \ldots, y_m) = (y', y'')$$

and then

$$F(y) = (f(y), y'')$$

for $y \in O$. The Jacobian matrix of F at a has the form

$$J_F(a) = \begin{pmatrix} \frac{\partial f_1}{\partial y_1}(a) & \cdots & \frac{\partial f_1}{\partial y_n}(a) & \cdots & \frac{\partial f_1}{\partial y_m}(a) \\ \vdots & & \vdots & & \vdots \\ \frac{\partial f_n}{\partial y_1}(a) & \cdots & \frac{\partial f_n}{\partial y_n}(a) & \cdots & \frac{\partial f_n}{\partial y_m}(a) \\ & & & & \\ & 0_{m-n,n} & & I_{m-n} & \end{pmatrix}.$$

As $J_F(a)$ is invertible, from the inverse mapping theorem we know that there is an open neighbourhood U of a such that $F_{|U}$ is a C^1-diffeomorphism onto its image V. Restricting $F_{|u}$ if necessary, we may suppose that V is an open cube centred on $(f(a), a'')$. If we set $\alpha_1 = F_{|U}^{-1}$, then α_1 is a C^1-diffeomorphism. For $z \in V$ there exists $y \in U$ such that $F(y) = z$, i.e.,

$$z = (z_1, \ldots, z_m) = (f_1(y), \ldots, f_n(y), y_{n+1}, \ldots, y_m)$$

and so

$$f \circ \alpha_1(z) = f(y) = (f_1(y), \ldots, f_n(y)) = (z_1, \ldots, z_n).$$

We now define α_2 on $\mathbb{R}^m$ to be the translation by $(f(a), a'')$ and β on $\mathbb{R}^n$ to be the translation by $-f(a)$. In addition, we set $U' = \alpha_2^{-1}(V)$; clearly U' is an open cube centred on the origin. Now setting $\alpha = \alpha_1 \circ \alpha_2$ we obtain a C^1-diffeomorphism from U' onto U and

$$\beta \circ f \circ \alpha(x_1, \ldots, x_n, \ldots, x_m) = (x_1, \ldots, x_n)$$

for all $x \in U'$. $\square$

We now turn to immersions.

Proposition 8.3. *Let O be an open subset of $\mathbb{R}^m$, $f : O \longrightarrow \mathbb{R}^n$ a C^1-immersion and $a \in O$. There exist open neighbourhoods U of a and V of $f(a)$, an open cube U' in $\mathbb{R}^m$, centred on the origin, and C^1-diffeomorphisms $\alpha : U' \longrightarrow U$ and $\beta : V \longrightarrow V' \subset \mathbb{R}^n$ such that*

$$\beta \circ f \circ \alpha(x_1, \ldots, x_m) = (x_1, \ldots, x_m, 0, \ldots, 0)$$

for all $x \in U'$.

Proof. The Jacobian matrix $J_f(a)$ has m independent rows; without loss of generality, let us suppose that these are the first m rows. We now consider the mapping

$$F : O \times \mathbb{R}^{n-m} \longrightarrow \mathbb{R}^n, (x, w) \longmapsto f(x) + (0, w).$$

The Jacobian matrix of F at $(a, 0)$ can be written

$$J_F(a, 0) = \begin{pmatrix} \frac{\partial f_1}{\partial x_1}(a) & \cdots & \frac{\partial f_1}{\partial x_m}(a) & & \\ \vdots & \vdots & \vdots & 0_{m, n-m} & \\ \frac{\partial f_m}{\partial x_1}(a) & \cdots & \frac{\partial f_m}{\partial x_m}(a) & & \\ \vdots & \vdots & \vdots & & I_{n-m} \\ \frac{\partial f_n}{\partial x_1}(a) & \cdots & \frac{\partial f_n}{\partial x_m}(a) & & \end{pmatrix}.$$

As $J_F(a, 0)$ is invertible, there is an open neighbourhood $\tilde{U}$ of $(a, 0)$ such that $F_{|\tilde{U}}$ is a C^1-diffeomorphism onto its image V. We can suppose that $\tilde{U}$ has the form $U \times \tilde{U}_2$, where U is an open neighbourhood of a in O and $\tilde{U}_2$ an open neighbourhood of 0 in $\mathbb{R}^{n-m}$. If we set $\beta_1 = F_{|\tilde{U}}^{-1}$, then β_1 is a C^1-diffeomorphism. For $z \in U$ and $y = f(z)$, we have $F(z, 0) = y$, therefore

$$\beta_1 \circ f(z) = \beta_1(y) = (z, 0),$$

or

$$\beta_1 \circ f(z_1, \ldots, z_m) = (z_1, \ldots, z_m, 0, \ldots, 0).$$

Reducing $\tilde{U}$ if necessary, we may suppose that U is an open cube centred on a. We now define α on $\mathbb{R}^m$ to be the translation by a and set $U' = \alpha^{-1}(U)$; clearly U' is an open cube centred on the origin. We also define β_2 on $\mathbb{R}^n$ to be the translation by $(-a, 0)$ and set $\beta = \beta_2 \circ \beta_1$. Then β is a C^1-diffeomorphism from V onto a subset V' of $\mathbb{R}^n$ and

$$\beta \circ f \circ \alpha(x_1, \ldots, x_m) = (x_1, \ldots, x_m, 0, \ldots, 0)$$

for all $x \in U'$.   □

Let us now consider the case where the rank of f is constant, but strictly inferior to m and n.

Proposition 8.4. *Let O be an open subset of $\mathbb{R}^m$, $f : O \longrightarrow \mathbb{R}^n$ a C^1-mapping of rank $r < \min(m, n)$ and $a \in O$. There exist open neighbourhoods U of a and V of $b = f(a)$, an open cube U' in $\mathbb{R}^m$, centred on the origin, and C^1-diffeomorphisms $\alpha : U' \longrightarrow U$ and $\beta : V \longrightarrow V' \subset \mathbb{R}^n$ such that*

$$\beta \circ f \circ \alpha(x_1, \ldots, x_m) = (x_1, \ldots, x_r, 0, \ldots, 0)$$

for all $x \in U'$.

Proof. Let us first suppose that $a = 0 \in \mathbb{R}^m$, $b = 0 \in \mathbb{R}^n$ and

$$f'(0)(y_1, \ldots, y_r, \ldots, y_m) = (y_1, \ldots, y_r, 0, \ldots, 0).$$

We will refer to these conditions as conditions (C). For $y \in O$ we set

$$W(y) = (f_1(y), \ldots, f_r(y), y_{r+1}, \ldots, y_m).$$

As $W'(0) = \mathrm{id}_{\mathbb{R}^m}$, there exist open neighbourhoods U and U' of $0 \in \mathbb{R}^m$ such that $W : U \longrightarrow U'$ is a C^1-diffeomorphism. We now set $\alpha = W^{-1}$. If $x \in U'$, then there exists $y \in U$ such that

$$x = (f_1(y), \ldots, f_r(y), y_{r+1}, \ldots, y_m)$$

and so

$$
\begin{aligned}
f \circ \alpha(x) = f(y) &= (f_1(y), \ldots, f_r(y), f_{r+1}(y), \ldots, f_n(y)) \\
&= (x_1, \ldots, x_r, \phi_{r+1}(x), \ldots, \phi_n(x)),
\end{aligned}
$$

where $\phi_i = f_i \circ \alpha$. The Jacobian matrix of $f \circ \alpha$ has the form

$$
J_{f \circ \alpha}(x) = \begin{pmatrix} I_r & 0_{r,m-r} \\ \frac{\partial \phi_{r+1}}{\partial x_1}(x) & \cdots & \frac{\partial \phi_{r+1}}{\partial x_m}(x) \\ \vdots & & \vdots \\ \frac{\partial \phi_n}{\partial x_1}(x) & \cdots & \frac{\partial \phi_n}{\partial x_m}(x) \end{pmatrix}.
$$

As α is a diffeomorphism, $f \circ \alpha$ has constant rank r. This implies that $\frac{\partial \phi_{r+i}}{\partial x_j}(x) = 0$ for $j > r$. Restricting α if necessary, we can suppose that U' is an open cube $C^m(0, \epsilon)$. The functions $\phi_{r+1}, \ldots, \phi_n$ are constant for fixed $x_1, \ldots, x_r$ (Theorem 3.3). If, for $(x_1, \ldots, x_r) \in C^r(0, \epsilon) \subset \mathbb{R}^r$, we set

$$
\bar{\phi}_i(x_1, \ldots, x_r) = \phi_i(x_1, \ldots, x_r, 0, \ldots, 0),
$$

then $\bar{\phi}_i$ is of class C^1 and

$$
\phi_i(x_1, \ldots, x_r, x_{r+1}, \ldots, x_m) = \bar{\phi}_i(x_1, \ldots, x_r)
$$

for all $(x_1, \ldots, x_m) \in U'$. Now let us set

$$
\beta(x_1, \ldots, x_n) = (x_1, \ldots, x_r, x_{r+1} - \bar{\phi}_{r+1}(x_1, \ldots, x_r), \ldots, x_n - \bar{\phi}_n(x_1, \ldots, x_r))
$$

for $(x_1, \ldots, x_n)$ in the open cube $C^n(0, \epsilon)$. Then β is of class C^1 and the Jacobian matrix $J_\beta(0)$ is inferior triangular with all diagonal elements being equal to 1. Therefore there are open neighbourhoods V and V' of $0 \in \mathbb{R}^n$ such that $\beta : V \longrightarrow V'$ is a C^1-diffeomorphism. Restricting U' to a smaller cube if necessary, we may suppose that Im $f \circ \alpha \subset C^n(0, \epsilon)$. For $(x_1, \ldots, x_m) \in U'$ we have

$$
\begin{aligned}
\beta \circ f \circ \alpha(x_1, \ldots, x_m) &= \beta(x_1, \ldots, x_r, \phi_{r+1}(x_1, \ldots, x_m), \ldots, \phi_n(x_1, \ldots, x_m)) \\
&= \beta(x_1, \ldots, x_r, \bar{\phi}_{r+1}(x_1, \ldots, x_r), \ldots, \bar{\phi}_n(x_1, \ldots, x_r)) \\
&= (x_1, \ldots, x_r, 0, \ldots, 0).
\end{aligned}
$$

Let us now consider the general case. We define $\alpha_1 : \mathbb{R}^m \longrightarrow \mathbb{R}^m$ and $\beta_1 : \mathbb{R}^n \longrightarrow \mathbb{R}^n$ by

$$
\alpha_1(x) = x + a \qquad \text{and} \qquad \beta_1(y) = y - b.
$$

From Theorem 8.3 we know that there are matrices $A \in \mathcal{M}_m(\mathbb{R})$ and $B \in \mathcal{M}_n(\mathbb{R})$ such that

$$
B J_f(a) A = \begin{pmatrix} I_r & 0_{r,m-r} \\ 0_{n-r,r} & 0_{n-r,m-r} \end{pmatrix}.
$$

For $x \in \mathbb{R}^m$ and $y \in \mathbb{R}^n$ we now set

$$\alpha_2(x) = Ax \qquad \text{and} \qquad \beta_2(y) = By$$

and then

$$\tilde{f} = \beta_2 \circ \beta_1 \circ f \circ \alpha_1 \circ \alpha_2 = \tilde{\beta} \circ f \circ \tilde{\alpha}.$$

If $\tilde{O} = \tilde{\alpha}^{-1}(O)$, then $\tilde{f}$ is defined on $\tilde{O}$ and satisfies the conditions (C). Therefore there exist open neighbourhoods $\tilde{U}$ and $\tilde{V}$ of the origin in $\mathbb{R}^m$ and $\mathbb{R}^n$, an open cube U' in $\mathbb{R}^m$, centred on the origin, and C^1-diffeomorphisms $\alpha_3 : U' \longrightarrow \tilde{U}$ and $\beta_3 : \tilde{V} \longrightarrow V' \subset \mathbb{R}^n$ such that

$$\tilde{\beta}_3 \circ \tilde{f} \circ \tilde{\alpha}_3(x_1, \ldots, x_r, \ldots, x_m) = (x_1, \ldots, x_r, 0, \ldots, 0).$$

We now set $U = \tilde{\alpha}(\tilde{U})$, $V = \tilde{\beta}^{-1}(\tilde{V})$, $\alpha = \tilde{\alpha} \circ \alpha_3$ and $\beta = \beta_3 \circ \tilde{\beta}$. Then U is a neighbourhood of a, V a neighbourhood of b and $\alpha : U' \longrightarrow U$ and $\beta : V \longrightarrow V'$ are C^1-diffeomorphisms. As $\beta \circ f \circ \alpha = \beta_3 \circ \tilde{f} \circ \alpha_3$, we have

$$\beta \circ f \circ \alpha(x_1, \ldots, x_r, \ldots, x_m) = (x_1, \ldots, x_r, 0, \ldots, x_m)$$

for all $x \in U'$. This finishes the proof. $\square$

We now put the previous three propositions together to obtain the *rank theorem*.

Theorem 8.4. *Let E and F be finite-dimensional normed vector spaces of respective dimensions m and n, O an open subset of E, $f : O \longrightarrow F$ a C^1-mapping of rank r and $a \in O$. Then there exist open neighbourhoods U of a and V of $b = f(a)$ and C^1-diffeomorphisms $\alpha : U' \longrightarrow U$ and $\beta : V \longrightarrow V'$, where U' is an open cube in $\mathbb{R}^m$, centred on the origin, such that*

$$\beta \circ f \circ \alpha(x_1, \ldots, x_r, \ldots, x_m) = (x_1, \ldots, x_r, 0, \ldots, 0)$$

for all $x \in U'$.

Proof. For the case where $E = \mathbb{R}^m$ and $F = \mathbb{R}^n$ we have already proved the result (Propositions 8.2, 8.3 and 8.4). For the general case it is sufficient to notice that E and $\mathbb{R}^m$ are isomorphic as are F and $\mathbb{R}^n$. $\square$

Exercise 8.5. Let E and F be normed vector spaces of respective dimensions m and n, O an open subset of E and f a C^1-mapping from O into F. Suppose that $a \in O$ with rk $f'(a) = m$ (resp. rk $f'(a) = n$). Show that there is a neighbourhood O' of a such that $f : O' \longrightarrow F$ is an immersion (resp. submersion).

8.4 Constrained Extrema

We have already looked at the problem of finding extrema of differentiable functions. Here we will return to the subject, but this time taking constraints into consideration. To be more precise, we consider a real-valued function f defined on some open subset O of a Banach space and look for possible local extrema a satisfying a condition of the form $g(a) = 0$, where g is a mapping from O into some Banach space F. In general, F is a euclidean space $\mathbb{R}^n$. Before arriving at a result characterizing such local extrema we will need to do some preliminary work.

Lemma 8.1. *Let E be a Banach space and l a linear mapping from E into $\mathbb{R}$. Then l is continuous if and only if the kernel of l is closed.*

Proof. If l is continuous, then $\mathrm{Ker}\, l$ is closed, because $\{0\}$ is closed.

Suppose now that $\mathrm{Ker}\, l$ is closed and let us write $K = \mathrm{Ker}\, l$. If $K = E$, then l is the zero mapping from E into $\mathbb{R}$ and so is continuous. If $K \neq E$, then there exists $a \in E \setminus K$. The distance from a to K, $d(a, K)$, is strictly positive, because K is closed. If $x \in E \setminus K$, then $l(x) \neq 0$ and $y = a - \frac{l(a)}{l(x)}x \in K$. Therefore

$$\frac{|l(a)|}{|l(x)|}\|x\| = \|y - a\| \geq d(a, K)$$

and it follows that

$$|l(x)| \leq \frac{|l(a)|}{d(a, K)}\|x\|.$$

As this is also true when $x \in K$, l is continuous and $\|l\| \leq \frac{|l(a)|}{d(a,K)}$. $\qquad\square$

Proposition 8.5. *Let E and F be Banach spaces, $l \in \mathcal{L}(E, F)$ surjective and $\phi \in E^*$. Suppose that $\mathrm{Ker}\, l$ has a closed complement L. Then the following two conditions are equivalent:*

(a) $\mathrm{Ker}\, l \subset \mathrm{Ker}\, \phi$;
(b) *There exists $\lambda \in F^*$ such that $\phi = \lambda \circ l$.*

Proof. The condition (b) clearly implies the condition (a).

Suppose now that the condition (a) is satisfied and let $y \in F$. Because l is surjective, $l^{-1}(y)$ is not empty. It is also closed, because l is continuous. If $x_1, x_2 \in l^{-1}(y)$, then $x_1 - x_2 \in \mathrm{Ker}\, l \subset \mathrm{Ker}\, \phi$, therefore $\phi(x_1) = \phi(x_2)$. If we define $\lambda(y)$ to be the common value of ϕ on the set $l^{-1}(y)$, then we obtain a real-valued function λ defined on F such that $\phi = \lambda \circ l$. The function λ is linear and continuous. Let us see why this is so.

First, the linearity. If $y = l(x)$, then $\lambda(y) = \phi(x)$ and for $\alpha \in \mathbb{R}$ we have

$$\alpha y = \alpha l(x) = l(\alpha x) \implies \lambda(\alpha y) = \phi(\alpha x) = \alpha\phi(x) = \alpha\lambda(y).$$

If $y_1 = l(x_1)$ and $y_2 = l(x_2)$, then

$$y_1 + y_2 = l(x_1) + l(x_2) = l(x_1 + x_2) \implies \lambda(y_1 + y_2) = \phi(x_1 + x_2)$$
$$= \phi(x_1) + \phi(x_2) = \lambda(y_1) + \lambda(y_2).$$

We have shown that λ is linear.

Now let us turn to the continuity. This is more difficult. The mapping $l_{|L}$ is linear continuous and bijective. From the corollary to the open mapping theorem (see Appendix 1), $l_{|L}$ is a normed vector space isomorphism. In addition, $\lambda(y) = 0$ if and only if $y = l(x)$, with $\phi(x) = 0$; however, this is the case if and only if $y \in l(\text{Ker}\,\phi)$. Therefore $\text{Ker}\,\lambda = l(\text{Ker}\,\phi)$. It is clear that

$$l_{|L}(L \cap \text{Ker}\,\phi) \subset l(\text{Ker}\,\phi).$$

Let $y \in l(\text{Ker}\,\phi)$ and $x \in E$ be such that $\phi(x) = 0$ and $y = l(x)$. We can write $x = x_1 + x_2$, with $x_1 \in \text{Ker}\,l$ and $x_2 \in L$. However, $\text{Ker}\,l \subset \text{Ker}\,\phi$ and so $\phi(x) = \phi(x_2)$. It follows that $x_2 \in L \cap \text{Ker}\,\phi$ and $y = l(x) = l(x_2)$. Hence

$$l(\text{Ker}\,\phi) \subset l_{|L}(L \cap \text{Ker}\,\phi).$$

To sum up, we have shown that

$$\text{Ker}\,\lambda = l_{|L}(L \cap \text{Ker}\,\phi).$$

As $l_{|L}$ is a normed vector space isomorphism and $L \cap \text{Ker}\,\phi$ is closed, $\text{Ker}\,\lambda$ is a closed subspace of F. From Lemma 8.1, λ is continuous.   □

Remark. The continuous linear mapping λ is unique. Suppose that there exists λ_1 satisfying the conditions of the proposition and that $\lambda_1 \neq \lambda$. Then there is a y such that $\lambda_1(y) \neq \lambda(y)$. If $y = l(x)$, then

$$\phi(x) = \lambda_1(l(x)) \neq \lambda(l(x)) = \phi(x),$$

a contradiction. Hence λ is unique.

We are now in a position to state and prove the result alluded to above concerning a relative minimum or maximum under constraints.

Theorem 8.5. *Let E and F be Banach spaces, O an open subset of E and $f : O \longrightarrow \mathbb{R}$ and $g : O \longrightarrow F$ mappings of class C^1. Suppose that $a \in A = g^{-1}(0)$ and that f has a relative extremum (minimum or maximum) at a. If $g'(a)$ is surjective and $\text{Ker}\,g'(a)$ has a closed complement L, then there is a unique $\lambda \in F^*$ such that $(f - \lambda \circ g)'(a) = 0$.*

Proof. If $\text{Ker}\,g'(a) \subset \text{Ker}\,f'(a)$, then from the previous proposition there is a unique element $\lambda \in F^*$ such that $f'(a) = \lambda \circ g'(a)$. However, $\lambda'(g(a)) = \lambda$, because $\lambda \in F^*$, and so we have $f'(a) = (\lambda \circ g)'(a)$. Therefore to prove the theorem it is sufficient to establish the inclusion $\operatorname{Ker} g'(a) \subset \operatorname{Ker} f'(a)$. Let us write K for $\operatorname{Ker} g'(a)$.

K is a closed subspace of E and so is a Banach space. The complementary subspace L is closed and so L too is a Banach space. The mapping

$$\phi : K \times L \longrightarrow E : (x, y) \longmapsto x + y$$

is a bijective continuous linear mapping and thus a normed vector space isomorphism. Let $\tilde{O} = \phi^{-1}(O)$ and $\tilde{g} = g \circ \phi$. Then $\tilde{g}$ is defined on $\tilde{O}$. Now, on the one hand

$$\tilde{g}'(x, y)(h, k) = \partial_1 \tilde{g}(x, y)h + \partial_2 \tilde{g}(x, y)k$$

and on the other

$$\tilde{g}'(x, y)(h, k) = g'(\phi(x, y)) \circ \phi'(x, y)(h, k) = g'(x + y)(h + k).$$

This implies that

$$\partial_1 \tilde{g}(x, y) = g'(x + y)_{|K} \qquad \text{and} \qquad \partial_2 \tilde{g}(x, y) = g'(x + y)_{|L}.$$

In particular, if $a = a_1 + a_2$ with $a_1 \in K$ and $a_2 \in L$, then

$$\partial_1 \tilde{g}(a_1, a_2) = g'(a)_{|K} \qquad \text{and} \qquad \partial_2 \tilde{g}(a_1, a_2) = g'(a)_{|L}.$$

It follows that $\partial_1 \tilde{g}(a_1, a_2) = 0$ and that $\partial_2 \tilde{g}(a_1, a_2)$ is a bijective continuous linear mapping from L onto F and so a normed vector space isomorphism.

We now apply the implicit mapping theorem: there exist an open neighbourhood $\tilde{O}'$ of (a_1, a_2), included in $\tilde{O}$, an open neighbourhood U of a_1 and a C^1-mapping $\psi : U \longrightarrow L$ such that the following statements are equivalent:

- $(x, y) \in \tilde{O}'$ and $\tilde{g}(x, y) = 0$;
- $x \in U$ and $y = \psi(x)$.

Next we consider the mapping

$$G : U \longrightarrow \mathbb{R}, x \longmapsto f \circ \phi(x, \psi(x)).$$

G has a relative extremum at a_1. As G is differentiable, we have $G'(a_1) = 0$. If we can show that $f'(a)k = G'(a_1)k$ for $k \in K$, then we are finished. Now, if $k \in K$, then

$$G'(a_1)k = f'(\phi(a_1, \psi(a_1))) \circ \phi'(a_1, \psi(a_1)) \circ (\operatorname{id}_K, \psi'(a_1))k$$
$$= f'(a)(\phi(k, \psi'(a_1)k))$$
$$= f'(a)(k + \psi'(a_1)k).$$

However, for $x \in U$ we have $\tilde{g}(x, \psi(x)) = 0$, which implies that

$$\partial_1 \tilde{g}(a_1, a_2)h + \partial_2 \tilde{g}(a_1, a_2) \circ \psi'(a_1)k = 0.$$

As $\partial_1 \tilde{g}(a_1, a_2) = 0$ and $\partial_2 \tilde{g}(a_1, a_2)$ is invertible, we must have $\psi'(a_1) = 0$. It follows that $f'(a)k = G'(a_1)k$ when $k \in K$. We have shown that $\operatorname{Ker} g'(a) \subset \operatorname{Ker} f'(a)$. This concludes the proof.   □

Remark. In Proposition 8.5 and the above theorem one of the conditions was the existence of a closed complementary subspace. In a Hilbert space or a finite-dimensional normed vector space this condition is not necessary, because it is always fulfilled.

   Suppose that $F = \mathbb{R}^n$. This means that the constraint g is composed of n real-valued functions: we have $g_i(x) = 0$ for $i = 1, \ldots, n$. If λ is a linear mapping from $\mathbb{R}^n$ into $\mathbb{R}$, then there is a unique set of constants $\lambda_1, \ldots, \lambda_n \in \mathbb{R}$ such that

$$\lambda(y_1, \ldots, y_n) = \lambda_1 y_1 + \cdots + \lambda_n y_n$$

for all $(y_1, \ldots, y_n) \in \mathbb{R}^n$. This means that the relation

$$(f - \lambda \circ g)'(a) = 0$$

can be written

$$f'(a) - (\lambda_1 g_1'(a) + \cdots + \lambda_n g_n'(a)) = 0.$$

If we further suppose that $E = \mathbb{R}^m$, then we can replace the above condition by the following:

$$\nabla f(a) - (\lambda_1 \nabla g_1(a) + \cdots + \lambda_n \nabla g_n(a)) = 0.$$

If we add the constraint equations $g_i(a) = 0$, then we see that a is a solution of a system of $m + n$ equations in $m + n$ unknowns. The constants λ_i are called *Lagrange multipliers.* Notice also that the surjectivity of $g'(a)$ is equivalent to the independence of the gradients $\nabla g_1(a), \ldots, \nabla g_n(a)$.

Example. Consider the function f defined on $\mathbb{R}^2$ by $f(x, y) = \cos x + \cos y$. On the domain

$$D = \left\{ (x, y) \in \mathbb{R}^2 : 0 \leq x \leq \frac{\pi}{2}, 0 \leq y \leq \frac{\pi}{2}, \sin x + \sin y = 1 \right\}.$$

f has both a maximum and a minimum, because f is continuous and D compact. Let (a, b) be a maximum and suppose that (a, b) lies in the set

$$D_1 = \left\{ (x, y) \in \mathbb{R}^2 : 0 < x < \frac{\pi}{2}, 0 < y < \frac{\pi}{2}, \sin x + \sin y = 1 \right\}.$$

Let us set $g(x, y) = \sin x + \sin y - 1$. Then

$$\nabla f(x, y) = (-\sin x, -\sin y) \qquad \text{and} \qquad \nabla g(x, y) = (\cos x, \cos y).$$

As $\nabla g \neq 0$ on D_1, there exists $\lambda \in \mathbb{R}$ such that

$$-\sin a = \lambda \cos a \qquad \text{and} \qquad -\sin b = \lambda \cos b,$$

which implies that

$$\tan a = \tan b = -\lambda.$$

Now, $a, b \in \left(0, \frac{\pi}{2}\right)$ and so $a = b$. It follows that $\sin a = \sin b$. However, $\sin a + \sin b = 1$ and so $\sin a = \sin b = \frac{1}{2}$; hence $a = b = \frac{\pi}{6}$. If $\left(\frac{\pi}{6}, \frac{\pi}{6}\right)$ is not a maximum, then the only other possibilities are $\left(0, \frac{\pi}{2}\right)$ or $\left(\frac{\pi}{2}, 0\right)$. However,

$$f\left(\frac{\pi}{6}, \frac{\pi}{6}\right) = \sqrt{3} > 1 = f\left(0, \frac{\pi}{2}\right) = f\left(\frac{\pi}{2}, 0\right)$$

and so $\left(\frac{\pi}{6}, \frac{\pi}{6}\right)$ is the unique maximum.

Let us now consider the minimum. If such a point were to lie in D_1, then it could only be the point $\left(\frac{\pi}{6}, \frac{\pi}{6}\right)$, which is not possible. It follows that there are two minima, namely $\left(0, \frac{\pi}{2}\right)$ and $\left(\frac{\pi}{2}, 0\right)$.

We now give an example with two constraints.

Example. We consider the intersection E of the surfaces

$$C : x^2 + y^2 = 5 \qquad \text{and} \qquad P : x + 2y + z = 0.$$

We aim to find those points in E which are closest and those which are furthest from the origin. First we set

$$f(x, y, z) = x^2 + y^2 + z^2, \quad g_1(x, y, z) = x^2 + y^2 - 5 \quad \text{and} \quad g_2(x, y, z) = x + 2y + z.$$

The point (a, b, c) minimizes (resp. maximizes) the distance to the origin if and only if (a, b, c) is a minimum (resp. maximum) of f. E is closed and bounded, hence compact. As f is continuous, f has a maximum and a minimum on E. Now,

$$\nabla g_1(x, y, z) = (2x, 2y, 0) \qquad \text{and} \qquad \nabla g_2(x, y, z) = (1, 2, 1),$$

which are linearly dependent if and only if $x = y = 0$. However, if this is so, then $g_1(x, y, z) = -5 \neq 0$, which means that $(x, y, z) \notin E$. Therefore, if (a, b, c) minimizes (or maximizes) f on E, then there exist $\lambda_1, \lambda_2 \in \mathbb{R}$ such that

$$\nabla f(a, b, c) - \lambda_1 \nabla_1 g_1(a, b, c) - \lambda_2 \nabla_1 g_2(a, b, c) = 0.$$

This leads us to look for the solutions $(S, \Lambda) \in \mathbb{R}^3 \times \mathbb{R}^2$ of the following system of equations:

$$2x - 2\lambda_1 x - \lambda_2 = 0$$
$$2y - 2\lambda_1 y - 2\lambda_2 = 0$$
$$2z - \lambda_2 = 0$$
$$x^2 + y^2 - 5 = 0$$
$$x + 2y + z = 0.$$

This system has four solutions, namely

$$(S_1, \Lambda_1) = (-2, 1, 0, 1, 0), \quad (S_2, \Lambda_2) = (2, -1, 0, 1, 0),$$

$$(S_3, \Lambda_3) = (1, 2 - 5, 6, -10) \quad \text{and} \quad (S_4, \Lambda_4) = (-1, -2, 5, 6, 10).$$

We find that $f(S_1) = f(S_2) = 5$ and that $f(S_3) = f(S_4) = 30$. It follows that S_1 and S_2 minimize the distance to the origin and S_3 and S_4 maximize the distance.

Remark. In looking for constrained extrema we are often led to solving large systems of polynomials in several variables. Gröbner bases can be very useful in handling such problems. A good introduction to the subject is the book by Cox, Little and O'Shea [8].

Exercise 8.6. Show that the function $f : \mathbb{R}^3 \longrightarrow \mathbb{R}, (x, y, z) \longmapsto xyz$ has a maximum on the domain

$$D = \{(x, y, z) \in \mathbb{R}_+^3 : x^2 + 4y^2 + z = 4\}.$$

Find the points where f attains a maximum value.

Exercise 8.7. Find the points on the surface

$$S : (x - y)^2 - z^2 = 1$$

which minimize the distance to the origin.

Exercise 8.8. Consider the function f defined on $\mathbb{R}^n$ by

$$f(x_1, \ldots, x_n) = \sum_{1 \leq i < j \leq n} x_i x_j.$$

Show that f has a maximum and a minimum on the domain

$$D = \left\{ (x_1,\dots,x_n) \in \mathbb{R}^n : x_1 \geq 0, \dots, x_n \geq 0, \sum_{i=1}^{n} x_i = 1 \right\}.$$

Find the extrema.

Exercise 8.9. Let $x_1,\dots,x_n \in \mathbb{R}_+^*$ and set

$$A_n = \left\{ (u_1,\dots,u_n) \in \mathbb{R}^n : u_1 > 0, \dots, u_n > 0, \prod_{i=1}^{n} u_i = 1 \right\}.$$

Show that the function

$$f : \mathbb{R}^n \longrightarrow \mathbb{R}, (u_1,\dots,u_n) \longmapsto \frac{1}{n} \sum_{i=1}^{n} u_i x_i$$

has a unique minimum on A_n and that the minimum value of f is $(\prod_{i=1}^{n} x_i)^{\frac{1}{n}}$. Deduce the inequality

$$\left(\prod_{i=1}^{n} x_i \right)^{\frac{1}{n}} + \left(\prod_{i=1}^{n} y_i \right)^{\frac{1}{n}} \leq \left(\prod_{i=1}^{n} (x_i + y_i) \right)^{\frac{1}{n}}$$

for all $x_1,\dots,x_n, y_1,\dots,y_n \in \mathbb{R}_+^*$.

In the second example above, which is relatively simple, we obtained a system of five polynomial equations in five unknowns. With a little work the solutions of this system could be found. However, in general it is no easy matter to solve the system of equations found and, even once we have the solutions, to decide whether a given solution is a maximum, minimum or neither one nor the other. Nevertheless, for one sort of constrained optimization problem we can find a complete solution. Let $A \in \mathcal{M}_n(\mathbb{R})$ be symmetric and positive definite, $b \in \mathbb{R}^n$, $C \in \mathcal{M}_{mn}(\mathbb{R})$, with $m \leq n$ and $\mathrm{rk}\, C = m$, and $d \in \mathbb{R}^m$. We consider the problem of minimizing the real-valued function

$$f(x) = \frac{1}{2} x^t A x - b^t x$$

under the constraint(s)

$$Cx = d.$$

The function f is strictly convex on the convex set $S = \{x \in \mathbb{R}^n : Cx = d\}$. If λ is the smallest eigenvalue of the matrix A, then

$$f(x) = \frac{1}{2} x^t A x - b^t x \geq \frac{1}{2} \lambda \|x\|_2^2 - \|b\|_2 \|x\|_2$$

and so $f(x)$ approaches ∞ when $\|x\|_2$ approaches ∞. As S is closed, f has a minimum $\bar{x}$ on S, which is unique, because f is strictly convex. We are minimizing f on $\mathbb{R}^n$, which is open, under the constraints

$$\phi_i(x) = \sum_{j=1}^{n} c_{ij} x_j - d_i$$

for $i = 1, \ldots, m$.

Now, $\nabla\phi_i(x) = (c_{i1}, \ldots, c_{in})$ and the rank of the matrix C is m, therefore the gradients are independent. It follows that there exist $\lambda_1, \ldots, \lambda_m \in \mathbb{R}$ such that

$$\nabla f(\bar{x}) - \sum_{i=1}^{m} \lambda_i \nabla\phi_i(\bar{x}) = 0.$$

Letting $\bar{\lambda}$ be the vector of Lagrange multipliers, i.e., $\bar{\lambda} = (\lambda_1, \ldots, \lambda_m)$, we can write these expressions in matrix form:

$$A\bar{x} - b - C^t\bar{\lambda} = 0.$$

Now let us consider the matrix $B = CA^{-1}C^t \in \mathcal{M}_m(\mathbb{R})$. First,

$$B^t = (C^t)^t (A^{-1})^t C^t = CA^{-1}C^t = B,$$

because the symmetry of A implies that of A^{-1}. Therefore B is symmetric. Also, if $y \in \mathbb{R}^m$, then we have

$$y^t(CA^{-1}C^t)y = (C^t y)^t A^{-1}(C^t y) \geq 0,$$

because the positive definiteness of A implies that of A^{-1}. In addition,

$$y^t(CA^{-1}C^t)y = 0 \Longrightarrow C^t y = 0,$$

which in turn implies that $y = 0$, because $\operatorname{rk} C^t = m$. Hence B is positive definite.

We can now determine $\bar{\lambda}$ and so $\bar{x}$. We have

$$A\bar{x} - b - C^t\bar{\lambda} = 0 \Longrightarrow C\bar{x} - CA^{-1}b - CA^{-1}C^t\bar{\lambda} = 0$$
$$\Longrightarrow d - CA^{-1}b - B\bar{\lambda} = 0$$
$$\Longrightarrow \bar{\lambda} = B^{-1}(d - CA^{-1}b)$$

and, substituting for $\bar{\lambda}$, we obtain

$$\bar{x} = A^{-1}b + A^{-1}C^t\bar{\lambda} = A^{-1}\big(b + C^t B^{-1}(d - CA^{-1}b)\big).$$

Appendix 1: Bijective Continuous Linear Mappings

If E and F are normed vector spaces and $f \in \mathcal{L}(E, F)$ is bijective, then there is no guarantee that the inverse f^{-1} is continuous. (Of course this is not a problem if E and F are finite-dimensional, because all linear mappings between such spaces are continuous.) We will show here that the inverse is continuous if E and F are Banach spaces. In so doing we will prove a little more, namely the open mapping theorem. We will begin with a lemma, often referred to as Baire's theorem. We need a definition. A subset S of a normed vector space E is *dense* in E if the closure $\bar{S} = E$. This is equivalent to saying that for any open ball $B(x, r)$ the intersection $B(x, r) \cap S$ is not empty.

Lemma 8.2. *(Baire's theorem) Let E be a Banach space and $(S_n)_{n=1}^{\infty}$ a sequence of closed subsets of E each with empty interior. Then the interior of the union of these subsets is also empty.*

Proof. Let us write O_n for the complement cS_n of S_n. Then O_n is open and dense in E. We set $O = \cap_{n=1}^{\infty} O_n$. It is sufficient to show that O is dense in E. Let $B(x, r)$ be an open ball. We choose $x_1 \in B(x, r) \cap O_1$ and $r_1 > 0$ such that

$$\bar{B}(x_1, r_1) \subset B(x, r) \cap O_1 \quad \text{and} \quad 0 < r_1 < \frac{r}{2}.$$

Next we choose $x_2 \in B(x_1, r_1) \cap O_2$ and $r_2 > 0$ such that

$$\bar{B}(x_2, r_2) \subset B(x_1, r_1) \cap O_2 \quad \text{and} \quad 0 < r_2 < \frac{r_1}{2}.$$

Continuing in the same way we obtain a sequence (x_n) such that $\|x_n - x_{n-1}\| < \frac{r}{2^n}$. Using the triangle inequality we now obtain for $k \geq 1$

$$\|x_{n+k} - x_n\| \leq \|x_{n+k} - x_{n+k-1}\| + \cdots + \|x_{n+1} - x_n\|$$
$$< \frac{r}{2^{n+k}} + \cdots + \frac{r}{2^{n+1}} < \frac{r}{2^n}.$$

Thus (x_n) is a Cauchy sequence and has a limit l. As $(x_n) \subset B(x_1, r_1)$, l must lie in $\bar{B}(x_1, r_1)$ and so in $B(x, r)$. However, $l \in \cap_{n=1}^{\infty} \bar{B}(x_n, r_n) \subset \cap_{n=1}^{\infty} O_n$ and so $B(x, r) \cap O \neq \emptyset$. As any open ball intersects O, O is dense in E. $\square$

Baire's theorem has many applications. It is fundamental in establishing the next result from which the open mapping follows. Also important in proving this result is Exercise 1.11, which we recall: If A is a subset of a normed vector space E and O is a nonempty open subset of $\bar{A}$, then $O \cap A \neq \emptyset$.

Lemma 8.3. *Let E and F be Banach spaces and $L \in \mathcal{L}(E, F)$ surjective. Then the image of an open ball centred on the origin in E contains an open ball centred on the origin in F.*

Proof. To simplify the notation, let us write B_r (resp. B'_r) for the open ball of radius r centred on the origin in E (resp. F). As

$$L(B_r) = L(rB_1) = rL(B_1),$$

it is sufficient to prove the result for $r = 1$. We will first show that $\overline{L(B_1)}$ contains an open ball B'_ϵ. As L is surjective, we have $F = \cup_{n=1}^\infty \overline{L(B_n)}$. By Baire's theorem there is an $m \in \mathbb{N}^*$ such that int $\overline{L(B_m)} \neq \emptyset$. From Exercise 1.11 there exists an element y_0 of this open set lying in $L(B_m)$. If $y_0 = L(x_0)$ and $x \in B_m$, then $x - x_0 \in B_{2m}$ and so $L(x - x_0) \in L(B_{2m})$. It follows that $L(B_m) - y_0 \subset L(B_{2m})$. Therefore

$$\overline{L(B_m)} - y_0 = \overline{L(B_m) - y_0} \subset \overline{L(B_{2m})}.$$

As int $(\overline{L(B_m)} - y_0) = $ int $\overline{L(B_m)} - y_0$, 0 is an interior point of $\overline{L(B_m)} - y_0$ and so of $\overline{L(B_{2m})}$. Also,

$$\overline{L(B_{2m})} = \overline{2mL(B_1)} = 2m\overline{L(B_1)},$$

and so 0 is an interior point of $2m\overline{L(B_1)}$ and thus of $\overline{L(B_1)}$: there exists an open ball B'_ϵ, centred on the origin in F, such that $B'_\epsilon \subset \overline{L(B_1)}$.

Our next step is to show that $B'_\epsilon \subset L(B_3)$. Let $y \in B'_\epsilon$. In order to show that $y \in L(B_3)$ we will establish the existence of sequences $(x_n) \subset E$ and $(y_n) \subset F$, with $y_n = L(x_n)$, such that

$$\|x_n\|_E < \frac{1}{2^{n-1}} \quad \text{and} \quad \left\| y - \sum_{i=1}^n y_i \right\|_F < \frac{\epsilon}{2^n}.$$

Let $B(y, r)$ be an open ball centred on y of radius $r < \frac{\epsilon}{2}$ and included in B'_ϵ. As $B(y, r) \subset \overline{L(B_1)}$, there exists $y_1 \subset B(y, r) \cap L(B_1)$ (Exercise 1.11). Therefore there exists $x_1 \in B_1$ such that $L(x_1) = y_1$. Thus we have the first members of the sequences (x_n) and (y_n). Suppose that we have constructed the sequences up to n. Now,

$$B'_{\frac{\epsilon}{2^n}} = \frac{1}{2^n} B'_\epsilon \subset \frac{1}{2^n}\overline{L(B_1)} = \overline{L\left(B_{\frac{1}{2^n}}\right)}.$$

Let us set $y' = y - \sum_{i=1}^n y_i$ and let $B(y', r')$ be an open ball centred on y' of radius $r' < \frac{\epsilon}{2^{n+1}}$ and included in $B'_{\frac{\epsilon}{2^n}}$. As $B(y', r') \subset \overline{L\left(B_{\frac{1}{2^n}}\right)}$, there exists $y_{n+1} \in$

$B(y', r') \cap L(B_{\frac{1}{2^n}})$. Because $y_{n+1} \in L\left(B_{\frac{1}{2^n}}\right)$, there exists $x_{n+1} \in B_{\frac{1}{2^n}}$ such that $L(x_{n+1}) = y_{n+1}$. Also, as $y_{n+1} \in B(y', r')$, we have

$$\|y' - y_{n+1}\|_F < \frac{\epsilon}{2^{n+1}} \quad \text{or} \quad \|y - \sum_{i=1}^{n+1} y_i\|_F < \frac{\epsilon}{2^{n+1}}.$$

Hence we can construct the sequences up to order $n + 1$. This establishes the existence of the sequences. If we set $s_n = \sum_{i=1}^{n} x_i$, then (s_n) is a Cauchy sequence and so has a limit x, with $\|x\| \leq 2 < 3$, and

$$L(x) = \lim L(s_n) = \lim \sum_{i=1}^{n} L(x_i) = \lim \sum_{i=1}^{n} y_i = y.$$

Therefore $y \in L(B_3)$ and so $B'_\epsilon \subset L(B_3)$.

However, showing that $B'_\epsilon \subset L(B_3)$ is equivalent to showing that $B'_{\frac{\epsilon}{3}} \subset L(B_1)$. This completes the proof. $\square$

Theorem 8.6. *(Open mapping theorem) If E and F are Banach spaces and $L : E \longrightarrow F$ is a surjective continuous linear mapping, then L is an open mapping, i.e., L maps open subsets of E to open subsets of F.*

Proof. Let $O \subset E$ be open and $y \in L(O)$. There is an $x \in O$ such that $L(x) = y$. As O is open, we can find an open ball $B(x, r)$ lying in O. From the lemma we know that there is an open ball B'_ϵ in F such that $B'_\epsilon \subset L(B_r)$. We have

$$B'(y, \epsilon) = y + B'_\epsilon \subset y + L(B_r) = L(x + B_r) = L(B(x, r)) \subset L(O).$$

This shows that $L(O)$ is open. $\square$

We are now in a position to establish what we set out to prove.

Corollary 8.2. *If E and F are Banach spaces and $L : E \longrightarrow F$ is a bijective continuous linear mapping, then the inverse of L is also continuous.*

Proof. From the theorem L is open. This implies that L^{-1} is continuous. $\square$

Appendix 2: Contractions

We recall the definition of a contraction. Let E be a normed vector space, $S \subset E$ and f a mapping from S into S. We say that f is a *contraction mapping* or a *contraction* if there is a constant $k \in [0, 1)$ such that

$$\|f(x) - f(y)\| \leq k\|x - y\|.$$

Any such k is called a *contraction factor*. The following result is known as the *contraction mapping theorem*.

Theorem 8.7. *Let E be a Banach space, S a closed subset of E and $f : S \longrightarrow S$ a contraction. Then f has a unique fixed point x. In addition, if $x_0 \in S$ and the sequence $(x_n)_{n=0}^{\infty}$ is defined recursively by the relation $x_{n+1} = f(x_n)$, then $\lim_{n \to \infty} x_n = x$.*

Proof. Let $x_0 \in S$. We define a sequence $(x_n)_{n=0}^{\infty}$ by the relation $x_{n+1} = f(x_n)$. If k is a contraction factor and $m > n$, then

$$\|x_m - x_n\| \leq \|x_m - x_{m-1}\| + \|x_{m-1} - x_{m-2}\| + \cdots + \|x_{n+1} - x_n\|.$$

However,

$$\|x_{s+1} - x_s\| \leq k\|x_s - x_{s-1}\| \leq k^2\|x_{s-1} - x_{s-2}\| \leq \cdots \leq k^s\|x_1 - x_0\|,$$

therefore

$$\|x_m - x_n\| \leq (k^{m-1} + k^{m-2} + \cdots + k^n)\|x_1 - x_0\| \leq \frac{k^n}{1-k}\|x_1 - x_0\|.$$

It follows that (x_n) is a Cauchy sequence and so has a limit x, which belongs to S because S is closed.

The sequence $(f(x_n))$ is a subsequence of (x_n) and f is continuous, therefore

$$x = \lim f(x_n) = f(x).$$

Thus x is a fixed point of f. If $f(y) = y$, then

$$\|x - y\| = \|f(x) - f(y)\| \leq k\|x - y\|,$$

which is possible only if $x = y$. Therefore the fixed point is unique.

Let $x_0' \in S$ and the sequence $(x_n')_{n=0}^{\infty}$ be defined recursively by the relation $x_{n+1}' = f(x_n')$. From what we have just seen, the sequence (x_n') converges to a fixed point x' of f. The uniqueness of the fixed point implies that $x' = x$. □

Chapter 9
Vector Fields

Let E be a normed vector space and O an open subset of E. A continuous mapping X from O into E is called a *vector field*. If X is of class C^k, with $k \geq 1$, then we refer to X as a vector field of class C^k. If I is an open interval of $\mathbb{R}$ and ϕ a differentiable mapping from I into O such that

$$\dot{\phi}(t) = X(\phi(t))$$

for all $t \in I$, then we say that ϕ is an *integral curve* of X or a solution of the first-order differential equation $\dot{x} = X(x)$. As X and ϕ are both continuous, so is $X \circ \phi$ and it follows that ϕ is of class C^1. An integral curve ϕ defined on an open interval I is a *maximal integral curve* if it cannot be extended to an integral curve defined on an open interval strictly containing I.

Exercise 9.1. Show that, if the vector field X is of class C^k, then an integral curve ϕ of X is of class C^{k+1}. Thus, if X is smooth, then so is any integral curve of X.

We recall the definition of a Lipschitz mapping. Let E and F be normed vector spaces, S a subset of E and f a mapping of S into F. If there is a constant $K \geq 0$ such that

$$\|f(x) - f(y)\|_F \leq K\|x - y\|_E$$

for all $x, y \in S$, then we say that f is Lipschitz (or K-Lipschitz) and that K is a Lipschitz constant for f. We can generalize this idea. We say that f is locally Lipschitz if every point $x \in S$ has an open neighbourhood U such that f restricted to $U \cap S$ is Lipschitz. Clearly a locally Lipschitz mapping is continuous. Suppose that O is open, f of class C^1 and $x \in O$. As f' is continuous, f' is bounded on some open ball B centred on x. From Corollary 3.2 we see that f is Lipschitz on B and it follows that f is locally Lipschitz. We will be particularly interested in Lipschitz and locally Lipschitz vector fields.

Remark. A continuous mapping may not be bounded on a bounded subset; however, a Lipschitz mapping is bounded on bounded subsets. On the other hand, a continuous mapping is locally bounded on any subset S of a normed vector space, i.e., every $x \in S$ has an open neighbourhood U such that f restricted to $U \cap S$ is bounded.

Notation. Let $a \in \mathbb{R}$. We will write T_a for the translation mapping of $\mathbb{R}$ into $\mathbb{R}$ defined by

$$T_a(t) = t - a.$$

The integral curves of a vector field have an important elementary property, namely they are invariant under translation:

Proposition 9.1. *Let $\phi : I \longrightarrow O$ be an integral curve of a vector field $X :$ $O \longrightarrow E$ and $a \in \mathbb{R}$. Then $\phi \circ T_a$ is an integral curve of X defined on $I + a$. In addition, $\phi \circ T_a$ is maximal if and only if ϕ is maximal.*

Proof. The mapping $\phi \circ T_a$ is clearly defined on $I + a$ and for $t \in I + a$ we have

$$\frac{\mathrm{d}}{\mathrm{d}t}(\phi \circ T_a)(t) = \frac{\mathrm{d}}{\mathrm{d}t}\phi(t - a) = X(\phi(t - a)) = X(\phi \circ T_a(t)).$$

Therefore $\phi \circ T_a$ is an integral curve.

Suppose that ϕ is maximal. If $\phi \circ T_a$ is not maximal, then $\phi \circ T_a$ can be extended to an integral curve ϕ_1 defined on an open interval I_1 strictly containing $I + a$. However, $\phi_1 \circ T_{-a}$ is an integral curve defined on the interval $I_1 - a$. As I is strictly contained in $I_1 - a$ and $\phi = \phi_1 \circ T_{-a}$ on I, we have a contradiction to the maximality of ϕ. It follows that $\phi \circ T_a$ is maximal.

Using an analogous argument we can show that, if $\phi \circ T_a$ is maximal, then so is ϕ. $\square$

We often need to consider so-called time-dependent vector fields. Suppose that E and O are as above, J an open interval of $\mathbb{R}$ and Y a continuous mapping from $J \times O$ into E. Then we say that Y is a *time-dependent vector field*. In this case, for each $t \in J$, the mapping $Y_t : x \longmapsto Y(t, x)$ is a vector field. An integral curve of Y is a differentiable mapping ϕ from an open interval $I \subset J$ into O such that

$$\dot{\phi}(t) = Y(t, \phi(t))$$

for $t \in I$. As $Y(t, \phi(t))$ is continuous, ϕ is of class C^1. An integral curve is maximal if it cannot be extended to an integral curve defined on an open interval strictly containing I.

With a time-dependent vector field we may associate a vector field in a natural way. If $Y : J \times O \longrightarrow E$ is a time-dependent vector field and for $t \in J$ and $x \in O$ we set

$$X(t, x) = (1, Y(t, x)),$$

then we obtain a vector field defined on the open subset $J \times O$ of $\mathbb{R} \times E$. As might be expected, there is a relation between the integral curves of Y and X.

Proposition 9.2. *The integral curves ψ of X are of the form $\tilde{\phi} \circ T_a$, where*

$$\tilde{\phi}(t) = (t, \phi(t))$$

and ϕ is an integral curve of Y. ψ is maximal if and only if ϕ is maximal.

Proof. If $\phi : I \longrightarrow O$ is an integral curve of Y and we set $\tilde{\phi}(t) = (t, \phi(t))$ for $t \in I$, then it is easy to see that $\tilde{\phi}$ is an integral curve of X. If $a \in \mathbb{R}$, then as above $\tilde{\phi} \circ T_a$ is an integral curve of X defined on $I + a$:

$$\tilde{\phi} \circ T_a(t) = (t - a, \phi(t - a)).$$

Suppose now that ψ is an integral curve of X defined on an open interval I. Then

$$\psi(t) = (\alpha(t), \phi(t)),$$

where $\alpha(t) \in J$, $\phi(t) \in O$ and

$$\frac{\mathrm{d}}{\mathrm{d}t}\psi(t) = \left(1, Y(\alpha(t), \phi(t))\right).$$

As $\frac{\mathrm{d}}{\mathrm{d}t}\alpha(t) = 1$, there exists $a \in \mathbb{R}$ such that $\alpha(t) = t - a$. Therefore

$$\frac{\mathrm{d}}{\mathrm{d}t}\psi(t) = \left(1, Y(t - a, \phi(t))\right) \implies \frac{\mathrm{d}}{\mathrm{d}t}\phi(t) = Y(t - a, \phi(t)).$$

If we set $\phi_1(t) = \phi \circ T_{-a}$, then for $t \in I - a$ we obtain

$$\frac{\mathrm{d}}{\mathrm{d}t}\phi_1(t) = \frac{\mathrm{d}}{\mathrm{d}t}\phi(t + a) = Y(t, \phi(t + a)) = Y(t, \phi_1(t)),$$

and so ϕ_1 is an integral curve of Y defined on $I - a$. However,

$$\psi(t) = (t - a, \phi(t)) = (t - a, \phi_1(t - a)),$$

i.e., $\psi = \tilde{\phi}_1 \circ T_a$. Hence an integral curve of X always has the form given in the statement of the proposition.

To close, we notice that it is not difficult to see that an integral curve of X is maximal if and only if it is derived from a maximal integral curve of Y. $\quad\square$

Up to now we have supposed that integral curves exist. In the next section we will show that this is in general the case.

9.1   Existence of Integral Curves

Let $X : O \longrightarrow E$ be a locally Lipschitz vector field and $x_0 \in O$. There exists a closed ball $\bar{B}(x_0, r) \subset O$, with $r > 0$, and a constant $K \geq 0$ such that X restricted to $\bar{B}(x_0, r)$ is K-Lipschitz. As X is Lipschitz on $\bar{B}(x_0, r)$, X is bounded on $\bar{B}(x_0, r)$. We set

$$M = \sup_{x \in \bar{B}(x_0,r)} \|X(x)\|.$$

Theorem 9.1. *If E is a Banach space, $t_0 \in \mathbb{R}$ and $\epsilon > 0$, with $\epsilon M < r$, then there is a unique integral curve ϕ defined on the interval $I_\epsilon = (t_0 - \epsilon, t_0 + \epsilon)$ satisfying the condition $\phi(t_0) = x_0$. The image of ϕ lies in the open ball $B(x_0, r)$.*

Proof. The set of continuous mappings from $\bar{I}_\epsilon$ into E, which we write $C(\bar{I}_\epsilon, E)$, is a vector space. If we set

$$\|\gamma\| = \sup_{t \in \bar{I}_\epsilon} \|\gamma(t)\|_E$$

for $\gamma \in C(\bar{I}_\epsilon, E)$, then $\| \cdot \|$ defines a norm and with this norm $C(\bar{I}_\epsilon, E)$ is a Banach space. The set

$$S = \{\gamma \in C(\bar{I}_\epsilon, E) : \mathrm{Im}\,(\gamma) \subset \bar{B}(x_0, r), \gamma(t_0) = x_0\}$$

is a nonempty closed subset of $C(\bar{I}_\epsilon, E)$. We define a mapping F from S into $C(\bar{I}_\epsilon, E)$ by

$$F(\gamma)(t) = x_0 + \int_{t_0}^{t} X(\gamma(s))\mathrm{d}s.$$

Clearly $F(\gamma)(t_0) = x_0$. Also,

$$\|F(\gamma)(t) - x_0\|_E \leq \left| \int_{t_0}^{t} \|X(\gamma(s))\|_E \mathrm{d}s \right| \leq M|t - t_0| < r. \tag{9.1}$$

Therefore F is a mapping from S into S.

We will now show that, if n is sufficiently large, then F^n is a contraction. Let $\gamma_1, \gamma_2 \in S$. We claim that for $n \in \mathbb{N}^*$

$$\|F^n(\gamma_1) - F^n(\gamma_2)\| \leq \frac{(K\epsilon)^n}{n!} \|\gamma_1 - \gamma_2\|.$$

To establish this we will prove by induction that

$$\|F^n(\gamma_1)(t) - F^n(\gamma_2)(t)\|_E \leq \frac{K^n|t - t_0|^n}{n!} \|\gamma_1 - \gamma_2\|.$$

For $n = 1$ we have

$$\|F(\gamma_1)(t) - F(\gamma_2)(t)\|_E \le |\int_{t_0}^{t} \|X(\gamma_1(s)) - X(\gamma_2(s))\|_E ds|$$

$$\le K|\int_{t_0}^{t} \|\gamma_1(s) - \gamma_2(s)\|_E ds|$$

$$\le K|t - t_0|\|\gamma_1 - \gamma_2\|.$$

Suppose now that the result is true for n and consider the case $n + 1$. Replacing γ_1 and γ_2 by $F^n(\gamma_1)$ and $F^n(\gamma_2)$ in the calculation above we obtain

$$\|F^{n+1}(\gamma_1)(t) - F^{n+1}(\gamma_2)(t)\|_E \le K|\int_{t_0}^{t} \|F^n(\gamma_1)(s) - F^n(\gamma_2)(s)\|_E ds|$$

$$\le K|\int_{t_0}^{t} \frac{K^n|s - t_0|^n}{n!}\|\gamma_1 - \gamma_2\|ds|$$

$$= \frac{K^{n+1}|t - t_0|^{n+1}}{(n + 1)!}\|\gamma_1 - \gamma_2\|.$$

Therefore the result is true for $n + 1$ and so for all $n \in \mathbb{N}^*$. It now follows that

$$\|F^n(\gamma_1) - F^n(\gamma_2)\| \le \frac{(K\epsilon)^n}{n!}\|\gamma_1 - \gamma_2\|.$$

For n sufficiently large $(K\epsilon)^n < n!$ and so F^n is a contraction.

We now fix an n such that F^n is a contraction. From Theorem 8.7 F^n has a unique fixed point ϕ and, for any $\gamma \in S$, $\lim_{k\to\infty} F^{nk}(\gamma) = \phi$. Thus

$$\phi = \lim_{k\to\infty} F^{nk}(F(\phi)) = \lim_{k\to\infty} F(F^{nk}(\phi)) = F(\phi)$$

and so ϕ is a fixed point of F. If $F(\gamma) = \gamma$, then $F^n(\gamma) = \gamma$ and so $\gamma = \phi$, i.e., the fixed point of F is unique. To recapitulate, we have

$$\phi(t) = x_0 + \int_{t_0}^{t} X(\phi(s))ds.$$

Restricting ϕ to I_ϵ we obtain an integral curve defined on I_ϵ and of course $\phi(t_0) = x_0$. From the inequality (9.1) we see that the image of ϕ lies in the open ball $B(x_0, r)$.

Suppose now that ψ is another integral curve defined on I_ϵ with $\psi(t_0) = x_0$. Then for $t \in I_\epsilon$ we have

$$\psi(t) = x_0 + \int_{t_0}^{t} X(\psi(s))ds.$$

From Corollary 3.2, for $t \in I_\epsilon$ and $\eta > 0$ sufficiently small, we may write

$$\|\psi(t+\eta) - \psi(t)\|_E \leq \sup_{s \in (t, t+\eta)} \|\dot\psi(s)\|_E |\eta| \leq M\eta,$$

because $\dot\psi(s) = X(\psi(s))$. This implies that we can extend ψ to a continuous mapping defined on $\bar I_\epsilon$. Then

$$\psi(t_0 + \epsilon) = \lim_{\alpha \to \epsilon} \psi(t_0 + \alpha) = x_0 + \lim_{\alpha \to \epsilon} \int_{t_0}^{t_0 + \alpha} X(\psi(s))ds = x_0 + \int_{t_0}^{t_0 + \epsilon} X(\psi(s))ds.$$

In the same way

$$\psi(t_0 - \epsilon) = x_0 + \int_{t_0 - \epsilon}^{t_0} X(\psi(s))ds$$

and it follows that ψ is a fixed point of F. Hence $\psi = \phi$ on $\bar I_\epsilon$ and so on I_ϵ. This ends the proof.   □

Exercise 9.2. Show that the relation

$$\|\psi(t+\eta) - \psi(t)\|_E \leq M\eta$$

proved above implies that ψ can be extended to a continuous mapping defined on $\bar I_\epsilon$.

From here on, when speaking of vector fields, we will assume that the normed vector space E is a Banach space and that all vector fields are locally Lipschitz.

We have established the existence of an integral curve ϕ satisfying the condition $\phi(t_0) = x_0$ and its uniqueness in a local sense. In fact, there are many integral curves satisfying the condition; however, as we will soon see, there is one and only one maximal integral curve satisfying the condition.

Proposition 9.3. *Let ϕ and ψ be integral curves of the vector field X defined on the same open interval I. If $t_0 \in I$ and $\phi(t_0) = \psi(t_0)$, then $\phi(t) = \psi(t)$ for all $t \in I$.*

Proof. From Theorem 9.1 there is an open interval $\tilde I \subset I$ containing t_0 such that $\phi(t) = \psi(t)$ for all $t \in \tilde I$. Let I_1 be the union of all such intervals. Then I_1 is an open interval containing t_0 such that $\phi(t) = \psi(t)$ for all $t \in I_1$. Suppose that $I_1 \neq I$. If $I = (a, b)$ and $I_1 = (\alpha, \beta)$, then $a < \alpha$ or $\beta < b$. Without loss of generality, suppose that $\beta < b$. Because ϕ and ψ are continuous, $\phi(\beta) = \psi(\beta)$. Applying Theorem 9.1 we obtain an open interval I_2 containing β such that $\phi(t) = \psi(t)$ for all $t \in I_2$. The interval $I_3 = I_1 \cup I_2$ is an open interval on which ϕ and ψ agree and I_1 is strictly included in I_3, a contradiction. It follows that $\phi = \psi$ on I.   □

Theorem 9.2. *Let $t_0 \in \mathbb{R}$ and $x_0 \in O$, the domain of the vector field X. Then there is a unique maximal integral curve ϕ with $\phi(t_0) = x_0$.*

Proof. From what we have seen above there exists an interval $\tilde{I}$ on which is defined an integral curve $\tilde{\phi}$ with $\tilde{\phi}(t_0) = x_0$. Let I be the union of all such open intervals. For $t \in I$ we take any one of these integral curves $\tilde{\phi}$ defined at t and set

$$\phi(t) = \tilde{\phi}(t).$$

From the preceding proposition $\phi(t)$ does not depend on the integral curve $\tilde{\phi}$ which we have chosen, so ϕ is well-defined. Clearly ϕ is maximal and $\phi(t_0) = x_0$. If $\psi : I_1 \longrightarrow O$ is another maximal integral curve and $\psi(t_0) = x_0$, then $I_1 \subset I$ and, because ψ is maximal, $I_1 = I$. Using the preceding proposition again, we see that $\psi = \phi$. This ends the proof. □

Remarks. 1. If ϕ is the unique maximal integral curve with $\phi(t_0) = x_0$ and ϕ is defined on the interval I, then the unique maximal integral curve ψ with $\psi(t_1) = x_0$ is defined on $I + (t_1 - t_0)$ and $\psi = \phi \circ T_{t_1-t_0}$.
2. If an integral curve ϕ satisfies the condition $\phi(t_0) = x_0$, then we often say that ϕ passes through x_0 at time t_0.

We now consider a class of vector fields which is particularly easy to study. Let E be a Banach space and $A \in \mathcal{L}(E)$. If we set

$$X(x) = Ax$$

for $x \in E$, then X is a smooth vector field defined on E. Such vector fields are called *linear vector fields*. As

$$\|A(x) - A(y)\| \le |A| \|x - y\|,$$

X is Lipschitz. Let us consider the mapping

$$\alpha : \mathbb{R} \longrightarrow \mathcal{L}(E), t \longmapsto \exp(tA).$$

Proposition 9.4. *We have*

$$\dot{\alpha}(t) = \exp(tA) \circ A = A \circ \exp(tA).$$

Proof. First we notice that

$$\exp((t + h)A) - \exp(tA) = \exp(tA) \circ \exp(hA) - \exp(tA)$$
$$= \exp(tA) \circ (\exp(hA) - \mathrm{id}_E)$$
$$= \exp(tA) \circ \left(\sum_{i=1}^{\infty} \frac{h^i A^i}{i!} \right).$$

Therefore

$$\frac{1}{h}\left(\exp(t+h)A - \exp(tA)\right) = \exp(tA) \circ \left(A + \sum_{i=2}^{\infty} \frac{h^{i-1}A^i}{i!}\right)$$

and so

$$\lim_{h \to 0} \frac{1}{h}(\exp(t+h)A - \exp(tA)) = \exp(tA) \circ A.$$

As A commutes with every element of the series for $\exp(tA)$, A commutes with $\exp(tA)$ and so we have

$$\dot{\alpha}(t) = \exp(tA) \circ A = A \circ \exp(tA).$$

This ends the proof.   □

Now let us take $x_0 \in E$ and consider the mapping

$$\phi : \mathbb{R} \longrightarrow E, t \longmapsto \exp(tA)x_0.$$

Theorem 9.3. ϕ *is the maximal integral curve of the linear vector field*

$$X(x) = Ax$$

with $\phi(0) = x_0$.

Proof. Clearly $\phi(0) = x_0$ and ϕ is defined on $\mathbb{R}$, so we only need to show that $\dot{\phi}(t) = X(\phi(t))$. However, from Proposition 9.4

$$\dot{\phi}(t) = \frac{\mathrm{d}}{\mathrm{d}t}(\alpha(t)x_0) = \left(\frac{\mathrm{d}}{\mathrm{d}t}\alpha(t)\right)x_0 = A \circ \exp(tA)x_0 = A\phi(t),$$

hence the result.   □

Remark. If $t_1 \neq 0$ and we require the maximal integral curve ψ with $\psi(t_1) = x_0$, then it is sufficient to set $\psi = \phi \circ T_{t_1}$.

All maximal integral curves of linear vector fields are defined on $\mathbb{R}$. This is in general not the case. It may even be so that no maximal integral curve is defined on $\mathbb{R}$. Consider the vector field

$$X : \mathbb{R} \longrightarrow \mathbb{R} : x \longmapsto 1 + x^2.$$

X is of class C^1 and so locally Lipschitz. The maximal integral curve ϕ which passes through the point x_0 at time $t = 0$ has the form

$$\phi(t) = \tan(t + \arctan x_0)$$

and this curve is defined on the interval $I = (-\frac{\pi}{2} - \arctan x_0, \frac{\pi}{2} - \arctan x_0)$. The maximal integral curve ψ which passes through x_0 at time $t = t_1$ has the form $\psi = \phi \circ T_{t_1}$. The domain of ψ is $I + t_0$. Therefore no maximal integral curve is defined on $\mathbb{R}$.

Let X be a vector field defined on an open subset O of a Banach space E. If the point x_0 is such that $X(x_0) = 0$, then we say that x_0 is a an *equilibrium point* . All other points are said to be *regular points*. If $t_0 \in \mathbb{R}$ and x_0 is a regular point, then the constant mapping

$$\phi : \mathbb{R} \longrightarrow O, t \longmapsto x_0$$

is the maximal integral curve passing through x_0 at time $t = t_0$.

We will often write $\Phi(t, t_0, x_0)$ for the maximal integral curve ϕ satisfying the condition $\phi(t_0) = x_0$ and I_{t_0, x_0} for the domain of ϕ. If O is the domain of the vector field and we set

$$D = \cup_{t_0 \in \mathbb{R}, x_0 \in O} I_{t_0, x_0} \times \{t_0\} \times \{x_0\},$$

then Φ defines a mapping from D into O. We will refer to Φ as the *flow* of the vector field X. (The term 'flow' is often used for a related concept, which we will introduce further on.)

Example. If E is a Banach space, $A \in \mathcal{L}(E)$ and X the linear vector field defined on E by

$$X(x) = Ax,$$

then $D = \mathbb{R} \times \mathbb{R} \times E$ and

$$\Phi(t, t_0, x_0) = \exp((t - t_0)A)x_0.$$

Clearly Φ is continuous on D. In general, the domain D of the flow Φ of a vector field defined on an open subset O of a Banach space E is an open subset of $\mathbb{R}^2 \times O$ and Φ is continuous. We will prove this a little further on.

We will now establish some elementary properties of the flow Φ.

Proposition 9.5. *Let $t_0 \in \mathbb{R}$ and $x_0 \in O$. Then*

1. $\Phi(t_0, t_0, x_0) = x_0$.
2. *If $t_1, t \in \mathbb{R}$ and $(t_1, t_0, x_0) \in D$, then $(t, t_0, x_0) \in D$ if and only if $(t, t_1, \Phi(t_1, t_0, x_0)) \in D$ and in this case*

$$\Phi(t, t_0, x_0) = \Phi(t, t_1, \Phi(t_1, t_0, x_0)).$$

3. *If $(t_1, t_0, x_0) \in D$, then $(t_0, t_1, \Phi(t_1, t_0, x_0)) \in D$ and*

$$\Phi(t_0, t_1, \Phi(t_1, t_0, x_0)) = x_0.$$

Proof. Property 1 follows directly from the definition of Φ. Suppose that $(t_1, t_0, x_0) \in D$ and let $y = \Phi(t_1, t_0, x_0)$. The maximal integral curve passing through y at time t_1 is the maximal integral curve passing through x_0 at time t_0. Properties 2 and 3 now follow.   □

Using the first remark after Theorem 9.2 we see that $(t, t_1, x_0) \in D$ if and only if $(t - (t_1 - t_0), t_0, x_0) \in D$ and

$$\Phi(t, t_1, x_0) = \Phi(t - (t_1 - t_0), t_0, x_0).$$

In particular, with $t_0 = 0$ we obtain

$$\Phi(t, t_1, x_0) = \Phi(t - t_1, 0, x_0).$$

Now, with the help of Proposition 9.5, we derive the following result: if $(t_1, 0, x_0) \in D$, then we have

$$(t, 0, x_0) \in D \iff (t, t_1, \Phi(t_1, 0, x_0)) \in D \iff (t - t_1, 0, \Phi(t_1, 0, x_0)) \in D,$$

and in this case

$$\Phi(t, 0, x_0) = \Phi(t, t_1, \Phi(t_1, 0, x_0)) = \Phi(t - t_1, 0, \Phi(t_1, 0, x_0)).$$

We will come back to this property later.

Let us now return to time-dependent vector fields. Here we obtain analogous results to those which we proved above for vector fields. We need a definition. Let J be an open interval of $\mathbb{R}$, O an open subset of a normed vector space E and $Y(t, x)$ a time-dependent vector field defined on $J \times O$. We say that Y is locally Lipschitz in its second variable, if for every pair $(t_0, x_0) \in J \times O$ there is an open interval $I \subset J$, centred on t_0, an open ball $B \subset O$, centred on x_0, and a constant $K \geq 0$ such that

$$\|Y(t, x_1) - Y(t, x_2)\| \leq K\|x_2 - x_2\|$$

for $t \in I$ and $x_1, x_2 \in B$. If Y is of class C^1, then Y is locally Lipschitz in its second variable, because Y' is continuous and so bounded on an open ball in $J \times O$.

Theorem 9.4. *Let E be a Banach space, J an open interval of $\mathbb{R}$, O an open subset of E and Y a time-dependent vector field defined on $J \times O$, which is locally Lipschitz in its second variable. If $t_0 \in J$ and $x_0 \in O$, then there is a unique maximal integral curve ϕ of Y such that $\phi(t_0) = x_0$.*

Proof. We first prove the existence of an integral curve ϕ such that $\phi(t_0) = x_0$. To do so we proceed as in the proof of Theorem 9.1. As Y is locally Lipschitz in its second variable, there is a closed interval $\bar{I} \subset J$ of length 2ϵ and centred on t_0, a closed ball $\bar{B} \subset O$ of radius $r > 0$ and centred on x_0, and a constant $K \geq 0$ such that

$$\|Y(t, x_1) - Y(t, x_2)\| \leq K\|x_2 - x_2\|$$

for $t \in \bar{I}$ and $x_1, x_2 \in \bar{B}$. Thus Y is bounded on $\bar{I} \times \bar{B}$. We let

$$M = \sup_{(t,x) \in \bar{I} \times \bar{B}} \|Y(t,x)\|.$$

Reducing $\bar{I}$ if necessary, we may suppose that $\epsilon M < r$. We define S as in the proof of Theorem 9.1 and a mapping F from S into $C(\bar{I}, E)$ by

$$F(\gamma)(t) = x_0 + \int_{t_0}^{t} Y(s, \gamma(s)) \mathrm{d}s.$$

Similar calculations to those used in the proof of the theorem just mentioned provide us with a unique integral curve ϕ, defined on the interior I of $\bar{I}$ such that $\phi(t_0) = x_0$. It is easy to see that the proofs of Proposition 9.3 and Theorem 9.2 are also applicable here, so we obtain a unique maximal integral curve ϕ such that $\phi(t_0) = x_0$. $\qquad\square$

To close this section we consider a special class of time-dependent vector fields, namely *nonautonomous linear vector fields*. Let E be a normed vector space, J an open interval of $\mathbb{R}$ and $A : t \longmapsto A(t)$ a continuous mapping from J into $\mathcal{L}(E)$. If we set

$$Y(t, x) = A(t)x$$

for $x \in E$, then we obtain a time-dependent vector field defined on $J \times E$, called a nonautonomous linear vector field. In addition,

$$\|Y(t, x_1) - Y(t, x_2)\| \leq |A(t)| \|x_2 - x_1\|$$

for all $t \in J$ and $x_1, x_2 \in E$. Any $t_0 \in J$ lies in a compact nontrivial interval $\bar{I}$ contained in J. As A is continuous, $A(t)$ is bounded on $\bar{I}$ and it follows that Y is locally Lipschitz in its second variable. Therefore the previous theorem is applicable.

Studies of linear vector fields, especially in the finite-dimensional case, may be found in many places. The book by Avez [1] handles this subject very well.

9.2 Initial Conditions

We have seen that if x_0 belongs to the domain of a vector field and $t_0 \in \mathbb{R}$, then there is a unique maximal integral curve ϕ with $\phi(t_0) = x_0$. We often refer to the pair (t_0, x_0) as a set of initial conditions and say that ϕ satisfies these initial conditions. If we have another set of initial conditions (s_0, y_0) and ψ is the maximal integral curve satisfying these initial conditions, then it is natural to ask questions about the relation between the two curves, in particular, when the initial conditions are close. We aim to now consider two such questions. Let ϕ and ψ be integral curves of a vector field which at a given time t_0 pass respectively through the points x_0 and y_0, i.e., $\phi(t_0) = x_0$ and $\psi(t_0) = y_0$. We will consider the following questions:

1. if the curves ϕ and ψ are both defined on the interval $[t_0, t_1]$, can we estimate the distance between $\phi(t_1)$ and $\psi(t_1)$ as a function of that between x_0 and y_0?
2. if the curve ϕ is defined on the interval $[t_0, t_1]$ and y_0 is sufficiently close to x_0, is the curve ψ defined on $[t_0, t_1]$?

To study these questions we introduce a preliminary result, which will also be useful later on.

Lemma 9.1 (Gronwell's Inequality). *Let f and g be continuous functions from the closed interval $I = [a, b]$ into $\mathbb{R}_+$. If there is a constant $c \geq 0$ such that for $t \in I$*

$$f(t) \leq c + \int_a^t f(s)g(s)ds,$$

then for $t \in I$, we have

$$f(t) \leq c \exp \int_a^t g(s)ds.$$

Proof. Setting

$$\alpha(t) = c + \int_a^t f(s)g(s)ds$$

we have

$$\dot{\alpha}(t) = f(t)g(t) \leq \alpha(t)g(t),$$

because $f(t) \leq \alpha(t)$ and $g(t) \geq 0$. If we now set

$$\beta(t) = \alpha(t) \exp \left(-\int_a^t g(s)ds \right),$$

then we obtain

$$\dot{\beta}(t) = \dot{\alpha}(t) \exp \left(-\int_a^t g(s)ds \right) - \alpha(t)g(t) \exp \left(-\int_a^t g(s)ds \right)$$

$$= (\dot{\alpha}(t) - \alpha(t)g(t)) \exp \left(-\int_a^t g(s)ds \right) \leq 0.$$

Therefore

$$\beta(t) \leq \beta(a) = \alpha(a) = c$$

and so

$$f(t) \leq \alpha(t) = \beta(t) \exp \left(\int_a^t g(s)ds \right) \leq c \exp \left(\int_a^t g(s)ds \right).$$

This finishes the proof.   □

Corollary 9.1. *If f is a continuous mapping from a closed interval $I = [a, b]$ into $\mathbb{R}_+$ and there are constants $c \geq 0$ and $k \geq 0$ such that for $t \in I$*

$$f(t) \leq c + \int_a^t kf(s)\mathrm{d}s,$$

then for $t \in I$

$$f(t) \leq ce^{k(t-a)}.$$

Exercise 9.3. Let f, g and I be as in Lemma 9.1 and suppose that there is a constant $c \geq 0$ such that for $t \in I$

$$f(t) \leq c + \int_t^b f(s)g(s)\mathrm{d}s.$$

Adapt the proof of Lemma 9.1 to show that for $t \in I$ we have

$$f(t) \leq c \exp \int_t^b g(s)\mathrm{d}s.$$

Deduce that if f is a continuous function from a closed interval $I = [a, b]$ into $\mathbb{R}_+$ and there are constants $c \geq 0$ and $k \geq 0$ such that for $t \in I$

$$f(t) \leq c + \int_t^b kf(s)\mathrm{d}s,$$

then for $t \in I$

$$f(t) \leq ce^{k(b-t)}.$$

Let us now turn to the first question we set at the beginning of the section. We have the following result.

Theorem 9.5. *Let O be an open subset of the Banach space E and X a K-Lipschitz vector field defined on O. If ϕ and ψ are integral curves defined on the interval $I = [t_0, t_1]$, $\phi(t_0) = x_0$ and $\psi(t_0) = y_0$, then*

$$\|\phi(t) - \psi(t)\| \leq \|x_0 - y_0\|e^{K(t-t_0)}$$

for all $t \in I$.

Proof. For $t \in I$ let us set

$$f(t) = \|\phi(t) - \psi(t)\|.$$

Then we have

$$\phi(t) - \psi(t) - \phi(t_0) + \psi(t_0) = \int_{t_0}^{t} X(\phi(s)) - X(\psi(s)) ds$$

and so

$$f(t) \leq f(t_0) + \int_{t_0}^{t} K f(s) ds.$$

From Corollary 9.1 we obtain

$$\|\phi(t) - \psi(t)\| = f(t) \leq f(t_0) e^{K(t-t_0)} = \|x_0 - y_0\| e^{K(t-t_0)}.$$

This ends the proof.   □

Corollary 9.2. *Under the conditions of the theorem, if ϕ and ψ are integral curves defined on the interval $I = [a, b]$, $t_0 \in I$, $\phi(t_0) = x_0$ and $\psi(t_0) = y_0$, then for all $t \in I$ we have*

$$\|\phi(t) - \psi(t)\| \leq \|x_0 - y_0\| e^{K|t-t_0|}.$$

Proof. We have already proved the result for $t \in [t_0, b]$. If $t \in [a, t_0)$ and

$$f(t) = \|\phi(t) - \psi(t)\|,$$

then we have

$$\phi(t) - \psi(t) - \phi(t_0) + \psi(t_0) = \int_{t_0}^{t} X(\phi(s)) - X(\psi(s)) ds$$

and

$$f(t) \leq f(t_0) + \int_{t}^{t_0} K f(s) ds.$$

Using Exercise 9.3 we obtain

$$\|\phi(t) - \psi(t)\| = f(t) \leq f(t_0) e^{K(t_0-t)} = \|x_0 - y_0\| e^{K|t-t_0|}.$$

This finishes the proof.   □

Suppose now that the vector field X is only locally Lipschitz. If x_0 belongs to the domain O of X, then there is an open neighbourhood U of x_0 and a constant $K \geq 0$ such that X restricted to U is K-Lipschitz. In this case we obtain a result analogous to the above result, providing that $y_0 \in U$.

We now turn to the second question set at the beginning of the section. We will need a preliminary result.

Proposition 9.6. *Let X be a vector field defined on an open subset O of a Banach space E, $x_0 \in O$, $B(x_0, r)$ an open ball centred on x_0 and lying in O and $t_0 \in \mathbb{R}$. Then there are strictly positive numbers $\bar{\epsilon} > 0$ and $\bar{r} > 0$ such that, for any y in the open ball $B(x_0, \bar{r})$ and t in the open interval $(t_0 - \bar{\epsilon}, t_0 + \bar{\epsilon})$, the integral curve $\Phi(s, t, y)$ is defined on $(t_0 - \bar{\epsilon}, t_0 + \bar{\epsilon})$. Also, the image of the integral curve on this interval is included in the ball $B(x_0, r)$.*

Proof. There is a closed ball $\bar{B}(x_0, \tilde{r})$, with $\tilde{r} > 0$, included in $B(x_0, r)$ and a constant $K \geq 0$ such that X is K-Lipschitz on $\bar{B}(x_0, \tilde{r})$. We set $M = \sup_{x \in \bar{B}(x_0, \tilde{r})} \|X(x)\|$. Let $\epsilon > 0$ be such that $\epsilon M < \tilde{r}$. Suppose that $y \in \bar{B}(x_0, \frac{\tilde{r}}{3})$ and $t \in (t_0 - \frac{\epsilon}{3}, t_0 + \frac{\epsilon}{3})$. As $\bar{B}(y, \frac{\tilde{r}}{3})$ is included in $\bar{B}(x_0, \tilde{r})$, X is K-Lipschitz on $\bar{B}(y, \frac{\tilde{r}}{3})$. If $N = \sup_{x \in \bar{B}(y, \frac{\tilde{r}}{3})} \|X(x)\|$, then

$$\frac{2\epsilon N}{3} \leq \frac{2\epsilon M}{3} < \frac{2}{3}\tilde{r}.$$

From Theorem 9.1 the integral curve $\Phi(s, t, y)$ is defined on the interval $(t - \frac{2\epsilon}{3}, t + \frac{2\epsilon}{3})$. However, the interval $(t_0 - \frac{\epsilon}{3}, t_0 + \frac{\epsilon}{3})$ is included in $(t - \frac{2\epsilon}{3}, t + \frac{2\epsilon}{3})$ and so $\Phi(s, t, y)$ is defined on $(t_0 - \frac{\epsilon}{3}, t_0 + \frac{\epsilon}{3})$. With $\bar{\epsilon} = \frac{\epsilon}{3}$ and $\bar{r} = \frac{\tilde{r}}{3}$ we obtain the principal result. To conclude, if $s \in (t_0 - \frac{\epsilon}{3}, t_0 + \frac{\epsilon}{3})$, then $s \in (t - \frac{2\epsilon}{3}, t + \frac{2\epsilon}{3})$ and so we have

$$\Phi(s, t, y) \in B\left(y, \frac{2\tilde{r}}{3}\right) \subset \bar{B}(x_0, \tilde{r}) \subset B(x_0, r).$$

This ends the proof. $\qquad\square$

We are now in a position to handle the second question which we asked at the beginning of the section.

Theorem 9.6. *Let X be a vector field defined on an open subset O of a Banach space E, $(a, x_0) \in \mathbb{R} \times O$ and suppose that the integral curve $\Phi(s, a, x_0)$ is defined on the interval $[a, b]$, with $a < b$. Then there is an open neighbourhood U of x_0 such that the integral curve $\Phi(s, a, y_0)$ is defined on the interval $[a, b]$, whenever $y_0 \in U$.*

Proof. To simplify the notation let us write $\phi(t)$ for $\Phi(t, a, x_0)$. By definition $\phi(a) = x_0$. We set

$$C = \{(s, \phi(s)) : s \in [a, b]\}$$

and

$$C_u = \{(s, \phi(s)) \in C : s \in [a, u]\}.$$

C (resp. C_u) is the graph of ϕ restricted to $[a, b]$ (resp. $[a, u)$). As C is the image of a continuous mapping defined on a compact interval, C is a compact subset of $\mathbb{R} \times O$. For each $t \in [a, b]$, let $B_t = B(\phi(t), r_t)$ be an open ball in O such that X is Lipschitz on B_t (with Lipschitz constant K_t). From Proposition 9.6 we know that there is an interval $I_t = (t - \epsilon_t, t + \epsilon_t)$ and an open ball $\bar{B}_t$ centred on $\phi(t)$ such that the integral curve $\Phi(s, t', y')$ is defined on I_t for all $(t', y') \in I_t \times \tilde{B}_t$ and has its image in B_t.

The sets $U_t = I_t \times \tilde{B}_t$ form an open cover of C. As C is compact, we may extract a finite subcover $U_{t_1}, \ldots, U_{t_n}$ from the cover. If necessary, we may eliminate certain members and so obtain a finite cover $\mathcal{U}$ which is minimal, i.e., no member of $\mathcal{U}$ can be removed without destroying the covering property. If $\mathcal{U}$ has only one member, then the result is elementary, so let us suppose that this is not the case. We may suppose that the sets U_{t_i} are ordered in the following way:

- $(a, \phi(a)) \in U_{t_1}$;
- if $C_u \subset \cup_{i=1}^k U_{t_i}$ and $(u, \phi(u)) \notin \cup_{i=1}^k U_{t_i}$, then $(u, \phi(u)) \in U_{t_{k+1}}$.

The second condition simply means that the addition of a set U_{t_i} extends the covering of the set C. We may find a sequence $(s_k)_{k=2}^n$ such that $\phi(s_k) \in \tilde{B}_{t_i} \cap \tilde{B}_{t_k}$, with $i < k$ and $s_{k-1} < s_k$. Clearly $\phi(s_k) \in \tilde{B}_{t_i} \cap \tilde{B}_{t_k}$.

At this point we will introduce a little more notation. We set

$$K = \max_i K_i \qquad \text{and} \qquad \epsilon = \max_i \epsilon_{t_i}.$$

As $(a, \phi(a)) \in U_{t_1}$, we may find an open ball B centred on x_0 and lying in $\tilde{B}_{t_1}$. Suppose now that $y_0 \in B$ and, to simplify the notation, let us write $\psi(t)$ for $\Phi(t, a, y_0)$. The integral curve ψ is defined on I_{t_1} and from Theorem 9.5 we have

$$\|\psi(t) - \phi(t)\| \leq \|y_0 - x_0\| e^{2K\epsilon}$$

for all $t \in I_{t_1}$, in particular for s_2. As $\phi(s_2) \in \tilde{B}_{t_1} \cap \tilde{B}_{t_2}$, reducing B if necessary, we obtain $\psi(s_2) \in \tilde{B}_{t_1} \cap \tilde{B}_{t_2}$. This means that the maximal integral curve ψ_2 passing through $\psi(s_2)$ at time $t = s_2$ is defined on I_{t_2}. However, $\psi_2(s_2) = \psi(s_2)$ and so $\psi_2 = \psi$ and it follows that ψ is defined on $I_{t_1} \cup I_{t_2}$. If $n = 2$, then we are finished.

Suppose that $n \geq 3$. We will now show that ψ is defined on I_{t_3} if y_0 is sufficiently close to x_0. As ψ is defined on I_{t_2}, we have

$$\|\psi(t) - \phi(t)\| \leq \|\psi(s_2) - \phi(s_2)\| e^{2K\epsilon} \leq \|y_0 - x_0\| e^{4K\epsilon}$$

for all $t \in I_{t_2}$. This is also true for $t \in I_{t_1}$ and so for $t \in I_{t_1} \cup I_{t_2}$. In particular, we have

$$\|\psi(s_3) - \phi(s_3)\| \leq \|y_0 - x_0\| e^{4K\epsilon}.$$

As $\phi(s_3) \in \tilde{B}_{t_i} \cap \tilde{B}_{t_3}$, with $i < 3$, reducing B again if necessary, we obtain $\psi(s_3) \in \tilde{B}_{t_i} \cap \tilde{B}_{t_3}$. This means that the maximal integral curve ψ_3 passing through $\psi(s_3)$ at time $t = s_3$ is defined on I_{t_3}. As $\psi_3(s_3) = \psi(s_3)$, we have $\psi_3 = \psi$. Therefore ψ is defined on $I_{t_1} \cup I_{t_2} \cup I_{t_3}$. Continuing in the same way we find that ψ is defined on $I_{t_1} \cup \cdots \cup I_{t_n}$ and so on $[a, b]$ for y_0 sufficiently close to x_0. $\qquad\square$

Corollary 9.3. *Let O be an open subset of a Banach space E, X a vector field defined on O, $x_0 \in O$, $I = [c, d]$ a closed interval of $\mathbb{R}$ and $t_0 \in I$. If the maximal* *integral curve ϕ passing through x_0 at time $t = t_0$ is defined on I, then there is an open ball B centred on x_0 such that the integral curve ψ passing through y_0 at time $t = t_0$ is defined on I when $y_0 \in B$.*

Proof. It is not difficult to modify the argument of Theorem 9.6 to the case where $(b, x_0) \in \mathbb{R} \times O$ and the integral curve $\Phi(s, b, x_0)$ is defined on the interval $[a, b]$. Considering in turn the intervals $[t_0, d]$ and $[c, t_0]$ we obtain the result. $\qquad\square$

Remark. We may also use the term initial conditions when referring to time-dependent vector fields. If ϕ is an integral curve of a time-dependent vector field and $\phi(t_0) = x_0$, then we say that ϕ satisfies the initial conditions (t_0, x_0).

9.3 Geometrical Properties of Integral Curves

Integral curves of vector fields have some interesting geometrical properties. Here will look at a few such properties.

Proposition 9.7. *If the images of two maximal integral curves have a point in common, then their images are the same.*

Proof. Let ϕ and ψ be maximal integral curves defined on intervals I and J and suppose that there exist points $t_1 \in I$ and $t_2 \in J$ such that $\phi(t_1) = \psi(t_2)$. From Proposition 9.1 $\psi_1 = \psi \circ T_{t_1 - t_2}$ is a maximal integral curve defined on $J_1 = J + (t_1 - t_2)$. Also,

$$\psi_1(t_1) = \psi(t_2) = \phi(t_1).$$

Given the uniqueness of maximal integral curves, $\psi_1 = \phi$. As ψ and ψ_1 have the same images, the images of ϕ and ψ are the same. $\qquad\square$

Exercise 9.4. What can we say of a maximal integral curve ϕ whose domain contains distinct points t_1 and t_2 such that $\phi(t_1) = \phi(t_2)$?

We have already observed that a maximal integral curve may not be defined on all of $\mathbb{R}$. If this is the case, then the curve has a particular behaviour when the variable approaches a finite bound of its domain.

Theorem 9.7. *Let X be a vector field defined on an open subset O of a Banach space E and ϕ a maximal integral curve whose domain is the interval $I = (a, b)$. If $b < \infty$ (resp. $a > -\infty$) and K is a compact subset of E, then there is real number $\epsilon > 0$ such that $\phi(t) \notin K$ when $t \in (b - \epsilon, b)$ (resp. $t \in (a, a + \epsilon)$).*

Proof. We will consider the case $b < \infty$. The other case can be handled in a similar way. Suppose that $b < \infty$ and that there is a compact set $K \subset O$ such that for any $\epsilon > 0$ there is a $t \in (b - \epsilon, b)$ with $\phi(t) \in K$. Then there is a sequence $(t_n) \subset (a, b)$ such that $\lim t_n = b$ and $\phi(t_n) \in K$ for all n. As K is compact, there is a subsequence (s_n) of the sequence (t_n) and $\bar{x} \in K$ such that $\lim \phi(s_n) = \bar{x}$. From Proposition 9.6 we know that there is an open ball $B = B(\bar{x}, r) \subset O$ and an interval $I = (b - \eta, b + \eta)$ such that the maximal integral curve $\Phi(s, t, x)$ is defined on I when $t \in I$ and $x \in B$. Let s_n' be such that $s_n' > b - \eta$ and $\phi(s_n') \in B$. Then the integral curve $\Phi(s, s_n', \phi(s_n'))$ is defined on I. However, this integral curve is the integral curve ϕ, which is not defined on the interval $[b, b + \eta)$, a contradiction. The result now follows. $\square$

Exercise 9.5. Suppose that the domain of the maximal integral curve ϕ is the interval (a, b), with $b < \infty$, and that $\lim_{t \to b} \phi(t) = x$. Show that x belongs to the boundary of the domain of the vector field.

9.4   Complete Vector Fields

If a vector field is such that all its maximal integral curves are defined on $\mathbb{R}$ then we say that it is *complete*. We have already seen that linear vector fields are complete. Clearly, if a vector field is complete, then the domain of its flow is $\mathbb{R}^2 \times O$, where O is the domain of the vector field. In this section we will explore complete vector fields a little further. However, before doing so, we will introduce another important notion.

Let E be a normed vector space, O an open subset of E and X a vector field defined on O. If f is a real-valued C^1-function defined on O such that, for any integral curve ϕ of X, $f \circ \phi$ is constant, then we say that ϕ is a *first integral* of X. Clearly, if f is a constant function, then f is a first integral; however, we are generally interested in nonconstant first integrals.

Proposition 9.8. *Let E be a Banach space, O an open subset of E and X a vector field defined on O. A C^1-function $f : O \longrightarrow \mathbb{R}$ is a first integral of X if and only if $f'(x)X(x) = 0$ for all $x \in O$.*

Proof. Let f be a first integral, $x \in O$ and ϕ an integral curve with $\phi(0) = x$. On an interval $I = (-\epsilon, +\epsilon)$, $f \circ \phi$ is constant, therefore

$$0 = \frac{\mathrm{d}}{\mathrm{d}t}(f \circ \phi)(t) = f'(\phi(t))\dot{\phi}(t) = f'(\phi(t))X(\phi(t)).$$

Setting $t = 0$ we obtain
$$f'(x)X(x) = 0.$$

Now suppose that
$$f'(x)X(x) = 0$$

for all $x \in O$. If ϕ is an integral curve defined on an open interval I, then

$$\frac{\mathrm{d}}{\mathrm{d}t}(f \circ \phi)(t) = f'(\phi(t))X(\phi(t)) = 0$$

for all $t \in I$. Therefore $f \circ \phi$ is constant on I and so f is a first integral of X. $\square$

Example. Consider the vector field X defined on $\mathbb{R}^3$ by

$$X(x, y, z) = (4xy - 1, -2x(1 + 2z), -2x(1 + 2y)).$$

If $f : \mathbb{R}^3 \longrightarrow \mathbb{R}$ is defined by

$$f(x, y, z) = y + y^2 - z - z^2,$$

then

$$\nabla f(x, y, z) = (0, 1 + 2y, -1 - 2z).$$

A simple calculation shows that

$$\nabla f(x, y, z) \cdot X(x, y, z) = 0$$

and so f is a first integral of X.

Remark. If we are working in a Hilbert space, then the condition of the theorem may be written

$$\langle \nabla f(x), X(x) \rangle = 0,$$

for all $x \in O$, i.e., f is a first integral if and only if the gradient of f is orthogonal to the vector field.

Exercise 9.6. Let O be an open subset of $\mathbb{R}^{2n}$ and f a real-valued C^1-function defined on O. If

$$J = \begin{bmatrix} 0 & I_n \\ -I_n & 0 \end{bmatrix}$$

and

$$X(x) = J \nabla f(x)$$

for $x \in O$, then X is called a Hamiltonian vector field. Show that f is a first integral of X.

Hamiltonian vector fields play an important role in mechanics and their study has given rise to other branches of mathematics, notably symplectic geometry. A good introduction to the subject may be found in the book by Hall and Meyer [13]. More advanced texts are those of Hofer and Zehnder [14] and Libermann, Marle and Tamboer [15].

First integrals give us useful information about vector fields. What interests us here is their role in determining whether a vector field is complete.

Theorem 9.8. *Suppose that a vector field X has a first integral f. If the inverse image $f^{-1}(c)$ is compact for all $c \in \mathbb{R}$, then the vector field X is complete.*

Proof. Let ϕ be a maximal integral curve defined on an open interval I. Then there exists an element $c \in \mathbb{R}$ such that $f \circ \phi(t) = c$ for all $t \in I$. Thus $\phi(t) \in f^{-1}(c)$

for all $t \in I$. However, $f^{-1}(c)$ is compact and so from Theorem 9.7 ϕ is defined on $\mathbb{R}$.   □

Exercise 9.7. Let X be the vector field defined on $\mathbb{R}^3$ by

$$X(x, y, z) = (y - z, z - x, x - y)$$

and f the real-valued function defined on $\mathbb{R}^3$ by

$$f(x, y, z) = x^2 + y^2 + z^2.$$

Show that f is a first integral of X, then that the vector field X is complete.

Using Gronwell's inequality, we may obtain another criterion for the completeness of a vector field.

Theorem 9.9. *If a vector field X is defined on an open subset of a finite-dimensional normed vector space E and is $(K$-$)$Lipschitz, then X is complete.*

Proof. Suppose that ϕ is a maximal integral curve with $\phi(t_0) = x_0$. Let $I = (a, b)$ be the domain of ϕ. If $b < \infty$, then for $t \in [t_0, b)$ we have

$$\|\phi(t) - x_0\| = \left\| \int_{t_0}^{t} X(\phi(s)) ds \right\|$$

$$\leq (t - t_0)\|X(x_0)\| + \int_{t_0}^{t} \|X(\phi(s)) - X(x_0)\| ds$$

$$\leq c + \int_{t_0}^{t} K\|\phi(s) - x_0\| ds,$$

where $c = (b - t_0)\|X(x_0)\|$. Applying Corollary 9.1 we obtain

$$\|\phi(t) - x_0\| \leq ce^{K(t - t_0)} \leq ce^{K(b - t_0)}.$$

This means that ϕ does not leave the compact set $\bar{B}(x_0, ce^{K(b-t_0)})$ when t approaches b, contradicting Theorem 9.7. Hence ϕ is defined on $[t_0, \infty)$. In the same way ϕ is defined on $(-\infty, t_0]$.   □

Suppose now that a vector field X is defined on an open subset O of a finite-dimensional normed vector space and that there is a compact subset K such that X vanishes outside of K. If $x_0 \in O \setminus K$, then $X(x_0) = 0$ and there is a unique maximal integral curve passing through x_0, namely the constant mapping

$$\phi : \mathbb{R} \longrightarrow O, t \longmapsto x_0.$$

Now let $x_0 \in K$ and suppose that the maximal integral curve ϕ satisfies the condition $\phi(t_0) = x_0$. If ϕ leaves K, then there exists $y \in O \setminus K$ lying in the image of ϕ. From what we have just seen, this implies that ϕ is constant, with $\phi(t) = y$ for all $t \in \mathbb{R}$, a contradiction to the fact that x_0 lies in its image. Therefore ϕ does not leave K and it follows that ϕ is defined on $\mathbb{R}$. Thus we have proved the following result:

Proposition 9.9. *If X is a vector field, defined on an open subset of a finite-dimensional normed vector space, which vanishes outside a compact subset included in its domain, then X is complete.*

In the appendix to this chapter it is proved that, if O is an open subset of a finite-dimensional normed vector space and K a compact subset of O, then there is a smooth real-valued function γ defined on O with compact support such that

- $\gamma(O) \subset [0, 1]$;
- $\gamma(x) = 1$ when $x \in K$.

Suppose now that X is a vector field of class C^1 defined on an open subset O of a finite-dimensional normed vector space E. Let K be any compact subset of O and γ as just defined. If we set, for $x \in O$,

$$\tilde{X}(x) = \gamma(x)X(x),$$

then $\tilde{X}$ is a composition of the C^1-mappings α and β, defined respectively on O and $\mathbb{R} \times E$, by

$$\alpha : x \longmapsto (\gamma(x), X(x))$$
$$\beta : (\lambda, x) \longmapsto \lambda x.$$

It follows that $\tilde{X}$ is a vector field of class C^1 vanishing outside a compact set. From the above proposition $\tilde{X}$ is complete. This shows that complete vector fields, at least on finite-dimensional normed vector spaces, are by no means rare.

Remark. In this section we have introduced the notion of first integral. We have seen that Hamiltonian systems always have an evident first integral. However, in general, it is not certain whether a given vector field has first integrals and, if so, how one may go about finding them. Interesting studies of this problem may be found in [11, 16, 18, 23].

Appendix: A Useful Result on Smooth Functions

In this section our aim is to show that, for any compact subset K of a finite-dimensional normed vector space E, it is possible to define a smooth real-valued function whose value is 1 on K. This result has many applications. To begin, we will prove an elementary result, which is also very useful.

Lemma 9.2. *Let $I = (a, b)$ be a bounded open interval of $\mathbb{R}$. Then there exists a smooth real-valued function g defined on $\mathbb{R}$ which is strictly positive on I and vanishes elsewhere.*

Proof. Let $f : \mathbb{R} \longrightarrow \mathbb{R}$ be defined as follows:

$$f(x) = \begin{cases} e^{-\frac{1}{x^2}} & x > 0 \\ 0 & \text{otherwise} \end{cases}.$$

From Taylor's theorem, for $x > 0$ we have

$$e^x = 1 + x + \frac{x^2}{2} + \cdots + \frac{x^n}{n!} + \frac{x^{n+1}}{(n+1)!} e^c,$$

where $c \in (0, x)$. This implies that $e^x > \frac{x^n}{n!}$ and we obtain

$$e^{\frac{1}{x^2}} > \frac{1}{n! x^{2n}} \implies e^{-\frac{1}{x^2}} < n! x^{2n} \implies \frac{1}{x^n} e^{-\frac{1}{x^2}} < n! x^n.$$

Therefore

$$\lim_{x \to 0+} \frac{1}{x^n} e^{-\frac{1}{x^2}} = 0.$$

Using this result it is not difficult to show, by induction, that f is smooth, with

$$f^{(n)}(x) = \begin{cases} \frac{P_n(x)}{x^{3n}} e^{-\frac{1}{x^2}} & x > 0 \\ 0 & \text{otherwise} \end{cases},$$

where $f^{(n)}(x)$ is the nth derivative of f at x and P_n a polynomial whose degree is $2n - 2$. If we set $g(x) = f(x - a) f(b - x)$, then g has the required properties. $\square$

Let us now set

$$h(x) = 1 - \frac{\int_a^x g(t) dt}{\int_a^b g(t) dt}.$$

Then h is smooth, $h(x) = 1$ if $x \leq a$, $h(x) = 0$ if $x \geq b$ and $h(x) \in (0, 1)$ if $x \in (a, b)$.

Suppose now that E is an n-dimensional normed vector space and that $l : E \longrightarrow$ $\mathbb{R}^n$ is a linear isomorphism. We define a scalar product on E by setting

$$\langle x, y \rangle = l(x) \cdot l(y)$$

for $x, y \in E$. We will write $\| \cdot \|$ for the associated norm.

Lemma 9.3. *Let* $\alpha, \beta \in \mathbb{R}_+^*$, *with* $\alpha < \beta$, *and* $x_0 \in E$. *If we set* $a = \alpha^2$ *and* $b = \beta^2$ *and define h as above, then the mapping*

$$u : E \longrightarrow \mathbb{R}, x \longmapsto h(\|x - x_0\|^2)$$

is smooth. Also, $u(x) = 1$ *if* $\|x - x_0\| \leq \alpha$, $u(x) = 0$ *if* $\|x - x_0\| \geq \beta$, *and* $u(x) \in (0, 1)$ *if* $\|x - x_0\| \in (\alpha, \beta)$.

Proof. The mapping $x \longmapsto \|x - x_0\|^2$ is smooth, because a scalar product is a continuous bilinear mapping and so smooth. As u is a composition of smooth mappings, u is smooth.

The proof of the second part of the lemma is elementary. $\square$

We will now prove the result referred to at the beginning of the appendix. We define the *support* of a real-valued function f defined on a subset S of a normed vector space E as follows:

$$\operatorname{supp} f = \overline{\{x \in S : f(x) \neq 0\}}.$$

Theorem 9.10. *Let E be a finite-dimensional normed vector space, O an open subset of E and K a compact subset of O. Then there is a smooth real-valued function γ defined on O, with compact support, such that $\gamma(O) \subset [0, 1]$ and $\gamma(x) = 1$ if $x \in K$.*

Proof. As norms on a finite-dimensional normed vector space are equivalent, a set is open, closed or compact, independently of the norm chosen. We will use the norm $\| \cdot \|$ on E derived from the scalar product defined above.

The complement cO of O is closed and $cO \cap K = \emptyset$. It follows that there is a real number $\epsilon > 0$ such that

$$\operatorname{dist}(K, cO) = \inf(\|x - y\| : x \in K, y \in cO) = \epsilon.$$

For each $a \in K$ the closed ball $\bar{B}(a, \frac{3\epsilon}{4})$ is contained in O. The open balls $B(a, \frac{\epsilon}{2})$, with $a \in K$, form an open cover of K. As K is compact, we may find a finite number of these balls, $B(a_1, \frac{\epsilon}{2}), \ldots, B(a_p, \frac{\epsilon}{2})$, which also cover K.

From Lemma 9.3, for each i there is a smooth real-valued function u_i, defined on E, such that $u_i(E) \subset [0, 1]$, $u_i(x) = 1$ for $x \in \bar{B}(a_i, \frac{\epsilon}{2})$ and $u_i(x) = 0$ for $x \notin B(a_i, \frac{3\epsilon}{4})$. If we set

$$\bar{\gamma}(x) = 1 - \prod_{i=1}^{p}(1 - u_i(x))$$

for $x \in E$, then

$$\mathrm{supp}\,\bar{\gamma} \subset \bigcup_{i=1}^{p} \bar{B}\left(a_i, \frac{3\epsilon}{4}\right),$$

a compact subset of O. The mapping $\gamma = \bar{\gamma}_{|O}$ has the required properties.   □

Chapter 10
The Flow of a Vector Field

In the last chapter we defined and studied some elementary properties of the flow of a vector field. In this chapter we will explore the flow in more detail. In general, normed vector spaces will be Banach spaces.

10.1 Continuity of the Flow

When discussing the flow in the previous chapter we spoke of its continuity. We will now prove this result.

Theorem 10.1. *The domain of the flow Φ of a vector field X defined on an open subset of a Banach space E is an open subset of $\mathbb{R}^2 \times E$ and the flow is continuous on its domain.*

Proof. Let D be the domain of Φ and $(s_0, t_0, x_0) \in D$. First suppose that $s_0 \geq t_0$. There exists $\epsilon > 0$ such that the maximal integral curve $\Phi(u, t_0, x_0)$ is defined on the interval $\bar{I} = [t_0 - \epsilon, s_0 + \epsilon]$. From Corollary 9.3 there is an open ball B centred on x_0 such that for all $x \in B$ the maximal integral curve $\Phi(u, t_0, x)$ is defined on the interval $\bar{I}$. Now let $t \in (t_0 - \frac{\epsilon}{2}, t_0 + \frac{\epsilon}{2})$, $s \in (s_0 - \frac{\epsilon}{2}, s_0 + \frac{\epsilon}{2})$ and $x \in B$. From Sect. 9.1 we see that

$$(s, t, x) \in D \iff (s + (t_0 - t), t_0, x) \in D.$$

Given the conditions on s and t,

$$t_0 - \epsilon < s + (t_0 - t) < s_0 + \epsilon$$

and so $(s, t, x) \in D$. We have shown that (s_0, t_0, x_0) has an open neighbourhood V such that, if $(s, t, x) \in V$, then $(s, t, x) \in D$. A similar argument can be used in the case where $s_0 < t_0$ and so it follows that D is an open subset of $\mathbb{R}^2 \times E$.

R. Coleman, *Calculus on Normed Vector Spaces*, Universitext,
DOI 10.1007/978-1-4614-3894-6_10, © Springer Science+Business Media New York 2012

We now consider the continuity. Reducing B if necessary, we may assume that the vector field X is Lipschitz on B (with Lipschitz constant K). Now, for $(s, t, x) \in V$ we have

$$\Phi(s, t, x) - \Phi(s_0, t_0, x_0) = \Phi(s, t, x) - \Phi(s, t, x_0) + \Phi(s, t, x_0) - \Phi(s_0, t_0, x_0).$$

We will call this equation (E). Notice first that

$$\|\Phi(s, t, x) - \Phi(s, t, x_0)\| \le \|x - x_0\| e^{K|s-t|}.$$

As $|s - t|$ is bounded, the difference of the first two terms is small when x is close to x_0. Now we observe that

$$\Phi(s, t, x_0) = \Phi(s + (t_0 - t), t_0, x_0).$$

Using the continuity of the maximal integral curve $\Phi(u, t_0, x_0)$ again, we see that the difference of the third and fourth terms of (E) is small when t is close to t_0 and s close to s_0. Hence, when (s, t, x) is close to (s_0, t_0, x_0), $\Phi(s, t, x)$ is close to $\Phi(s_0, t_0, x_0)$ and so Φ is continuous at (s_0, t_0, x_0). It follows that Φ is continuous on D. □

In the next sections we will see that the differentiability properties of the vector field carry over to the flow. This will involve more work than for the continuity.

10.2   Differentiability of the Flow

In this section we will show that the flow is of class C^1 if the vector field is of class C^1. Further on we will generalize this, showing that the flow is of class C^k if the vector field is of class C^k. In particular, smooth vector fields generate smooth flows. To prove that the flow is of class C^1, we will show that the partial differentials are defined and continuous. We will proceed by steps.

Proposition 10.1. *Let X be a vector field defined on an open subset of a Banach space E. If X is of class C^1, then the flow Φ of X has a partial differential $\partial_1 \Phi$, which is continuous on the domain D of Φ.*

Proof. Let $(s_0, t_0, x_0) \in D$. There is an open interval I (resp. J), centred on s_0 (resp. t_0) and an open ball B, centred on x_0, such that $V = I \times J \times B \subset D$. If we fix $(t, x) \in J \times B$, then the mapping $u \longmapsto \Phi(u, t, x)$ is an integral curve defined on I and so has a derivative for all $s \in I$. This implies that $\partial_1 \Phi(s, t, x)$ exists on V:

$$\partial_1 \Phi(s, t, x) w = w \frac{\partial}{\partial s} \Phi(s, t, x),$$

where $w \in \mathbb{R}$. As Φ and X are continuous and

$$\frac{\partial}{\partial s}\Phi(s,t,x) = X(\Phi(s,t,x)),$$

the mapping

$$(s,t,x) \longmapsto \frac{\partial}{\partial s}\Phi(s,t,x)$$

is continuous on V; hence $\partial_1 \Phi(s,t,x)$ is continuous on V. Therefore $\partial_1 \Phi$ is defined and continuous on D. $\qquad\qquad\square$

We now turn to the second differential of the flow.

Proposition 10.2. *If X is a vector field of class C^1 defined on an open subset of a Banach space E, then the flow Φ of X has a partial differential $\partial_2 \Phi$, which is continuous on the domain D of Φ.*

Proof. Let $(s_0, t_0, x_0) \in D$. As in the previous proposition, we take a neighbourhood $V = I \times J \times B \subset D$ of (s_0, t_0, x_0), with I and J open intervals, centred respectively on s_0 and t_0, and B an open ball centred on x_0. In Sect. 9.1 of the previous chapter, we saw that $(s,t,x) \in D$ if and only if $(s-t,0,x) \in D$ and in this case

$$\Phi(s,t,x) = \Phi(s-t,0,x).$$

If we fix $(s,x) \in I \times B$, then the mapping $t \longmapsto \Phi(s,t,x)$ is defined on J and at every point has a derivative:

$$\frac{\partial}{\partial t}\Phi(s,t,x) = \frac{\partial}{\partial t}\Phi(s-t,0,x) = -X(\Phi(s-t,0,x)) = -X(\Phi(s,t,x)).$$

It follows that $\partial_2 \Phi$ is defined on V:

$$\partial_2 \Phi(s,t,x)w = w\frac{\partial}{\partial t}\Phi(s,t,x),$$

where $w \in \mathbb{R}$. As X and Φ are continuous, $\partial_2 \Phi$ is continuous on V. Therefore $\partial_2 \Phi$ is defined and continuous on D. $\qquad\qquad\square$

Showing that $\partial_3 \Phi$ exists and is continuous is much more difficult.

Proposition 10.3. *If X is a vector field of class C^1 defined on an open subset O of a Banach space E, then the flow Φ of X has a partial differential $\partial_3 \Phi$, which is continuous on the domain D of Φ.*

Proof. We will divide the proof into two parts, namely the existence of $\partial_3 \Phi$ at every point of D, then the continuity of the mapping $\partial_3 \Phi$. In the first part we will obtain a characterization of $\partial_3 \Phi$, which we will use in the second part.

1. Existence For $s \in J_{(t,x)}$, the domain of the maximal integral curve $\Phi(s,t,x)$, we set

$$A_{(t,x)}(s) = X'(\Phi(s,t,x)) \in \mathcal{L}(E).$$

As X' and Φ are continuous, $A_{(t,x)}$ is continuous. We define a non-autonomous linear vector field on $\mathcal{L}(E)$ by

$$Y_{(t,x)}(s,L) = A_{(t,x)}(s) \circ L.$$

From our work in Sect. 9.1 of the previous chapter we know that, for any $s \in J_{(t,x)}$ and $L \in \mathcal{L}(E)$, there is a unique maximal integral curve satisfying the initial conditions (s,L) and that this curve is defined on $J_{(t,x)}$. Let $F(s,t,x)$ be the maximal integral curve of $Y_{(t,x)}$ satisfying the initial conditions (t, id_E). We will show that

$$F(s,t,x) = \partial_3 \Phi(s,t,x).$$

For $s \in J_{(t,x)}$ and h small, we set

$$\alpha(s, x+h) = \Phi(s,t,x+h) - \Phi(s,t,x) - F(s,t,x)h \in E.$$

As $F(t,t,x) = \mathrm{id}_E, \alpha(t, x+h) = 0$.

Now let us take $(s',t,x) \in D$, with $s' \geq t$. Notice that there is an open ball $B(0,r)$ such that $(s,t,x+h) \in D$ when $s \in [t,s']$ and $h \in B(0,r)$. If

$$\alpha(s', x+h) = o(h), \tag{10.1}$$

then indeed

$$F(s',t,x) = \partial_3 \Phi(s',t,x).$$

To prove (10.1) we will first show that for any $\epsilon > 0$ there is an $\eta > 0$ such that

$$\left\| \frac{\partial}{\partial s} \alpha(s, x+h) \right\| \leq \epsilon \| \Phi(s,t,x+h) - \Phi(s,t,x) \|$$

$$+ |X'(\Phi(s,t,x))| \|\alpha(s, x+h)\| \tag{10.2}$$

for all $s \in [t,s']$ and $\|h\| < \eta$. Taking the derivative of α with respect to s we obtain

$$\frac{\partial}{\partial s}\alpha(s, x+h) = \frac{\partial}{\partial s}\Phi(s,t,x+h) - \frac{\partial}{\partial s}(s,t,x) - \frac{\partial}{\partial s}F(s,t,x)h$$

$$= X(\Phi(s,t,x+h)) - X(\Phi(s,t,x)) - X'(\Phi(s,t,x)) \circ F(s,t,x)h$$

$$= X(\Phi(s,t,x+h)) - X(\Phi(s,t,x))$$

$$-X'(\Phi(s,t,x))(\Phi(s,t,x+h) - \Phi(s,t,x))$$

$$+X'(\Phi(s,t,x))\alpha(s, x+h).$$

To simplify the notation we will write $A(s, h)$ for the sum of the first three terms in the last expression. Using Corollary 3.3 we obtain

$$\|A(s,h)\| \leq \sup_{\lambda \in [0,1]} |X'(\lambda \Phi(s,t,x+h) + (1-\lambda)\Phi(s,t,x))$$

$$-X'(\Phi(s,t,x))| \|\Phi(s,t,x+h) - \Phi(s,t,x)\|.$$

Once again to simplify the notation, we will write $C(\lambda, s, h)$ for the expression

$$X'(\lambda \Phi(s,t,x+h) + (1-\lambda)\Phi(s,t,x)) - X'(\Phi(s,t,x)).$$

If we fix the variables other than h, then C converges to 0 when h converges to 0. We claim that for any $\epsilon > 0$ there is an $\eta \in (0, r]$ such that

$$\|h\| < \eta \Longrightarrow |C(\lambda, s, h)| < \epsilon.$$

If this is not the case, then we can find an $\epsilon > 0$ and sequences $(s_n) \subset [t, s']$, $(\lambda_n) \subset [0, 1]$ and $(h_n) \subset E$, with $\lim h_n = 0$, such that $|C(\lambda_n, s_n, h_n)| \geq \epsilon$. As the intervals $[t, s']$ and $[0, 1]$ are compact, we may suppose that the sequences (s_n) and (λ_n) have limits $\bar{s} \in [t, s']$ and $\bar{\lambda} \in [0, 1]$. Then

$$0 = |C(\bar{\lambda}, \bar{s}, 0)| = \lim |C(\lambda_n, s_n, h_n)| \geq \epsilon,$$

a contradiction. Therefore the claim is true. We may now write

$$\left\| \frac{\partial}{\partial s} \alpha(s, x+h) \right\| \leq \epsilon \|\Phi(s,t,x+h) - \Phi(s,t,x)\| + |X'(\Phi(s,t,x))| \|\alpha(s, x+h)\|$$

for all $s \in [t, s']$ and $\|h\| < \eta$. We have proved (10.2).

From Theorem 9.5 there is a constant $k > 0$ such that for h sufficiently small

$$\|\Phi(s,t,x+h) - \Phi(s,t,x)\| \leq k\|h\|$$

for all $s \in [t, s']$. We also notice that $|X'(\Phi(s,t,x))|$ is bounded by a constant $K \geq 0$ on the interval $[t, s']$. Therefore we finally have

$$\left\| \frac{\partial}{\partial s} \alpha(s, x+h) \right\| \leq \epsilon k\|h\| + K \|\alpha(s, x+h)\|.$$

As $\alpha(t, x+h) = 0$, we obtain

$$\|\alpha(s, x+h)\| \leq \epsilon(s-t)k\|h\| + K \int_t^s \|\alpha(u, x+h)\| du$$

$$\leq \epsilon(s'-t)k\|h\| + K \int_t^s \|\alpha(u, x+h)\| du.$$

An application of Corollary 9.1 gives us

$$\|\alpha(s, x + h)\| \leq \epsilon \|h\|(s' - t)ke^{K(s-t)}$$

and it follows that

$$\alpha(s, x + h) = o(h).$$

This proves that

$$\partial_3 \Phi(s, t, x) = F(s, t, x)$$

for $s \in [t, s']$ and, in particular, for s'.

Clearly, if $(s'', t, x) \in D$ with $s'' < t$, then we can apply an analogous argument to show that

$$\partial_3 \Phi(s'', t, x) = F(s'', t, x).$$

We have seen that $\partial_3 \Phi$ is defined on D and we have an explicit expression for this partial differential. Our next task is to show that this partial differential is continuous.

2. Continuity Let $(s_0, t_0, x_0) \in D$. For $\epsilon > 0$ and $r > 0$, let us set

$$\bar{I}_\epsilon = [s_0 - \epsilon, s_0 + \epsilon], \quad \bar{J}_\epsilon = [t_0 - \epsilon, t_0 + \epsilon] \quad \text{and} \quad B_r = B(x_0, r).$$

For ϵ and r sufficiently small, $\bar{I}_\epsilon \times \bar{J}_\epsilon \times B_r \subset D$. Let us suppose that $s_0 \geq t_0$. As the integral curve $\Phi(s, t_0, x_0)$ is defined on an open interval and is defined at s_0 and t_0, we may suppose that this curve is defined on the closed interval $\bar{K}_\epsilon = [t_0 - \epsilon, s_0 + \epsilon]$. There is an open ball B, centred on x_0, such that any maximal integral curve $\Phi(s, t_0, x)$ is defined on $\bar{K}_\epsilon$ if $x \in B$; we may suppose that $B_r = B$.

Suppose now that $t \in J_{\frac{\epsilon}{2}} = (t_0 - \frac{\epsilon}{2}, t_0 + \frac{\epsilon}{2})$ and $x \in B_r$. As

$$(s, t, x) \in D \iff (s + (t_0 - t), t_0, x) \in D,$$

the integral curve $\Phi(s, t, x)$ is defined on $\bar{K}_{\frac{\epsilon}{2}} = [t_0 - \frac{\epsilon}{2}, s_0 + \frac{\epsilon}{2}]$. Therefore any maximal integral curve of the non-autonomous linear vector field $Y_{(t,x)}$ is defined on $\bar{K}_{\frac{\epsilon}{2}}$. Above we wrote $F(s, t, x)$ for the maximal integral curve of $Y_{(t,x)}$ with initial conditions (t, id_E). To simplify the notation, here we will write $\psi(s)$ for $F(s, t, x)$ and $\phi(s)$ for the particular case $F(s, t_0, x_0)$.

Now, for $s \in \bar{K}_{\frac{\epsilon}{2}}$ we can write

$$\dot{\phi}(s) = \Big(X'(\Phi(s, t_0, x_0)) - X'(\Phi(s, t, x))\Big) \circ \phi(s) + X'(\Phi(s, t, x)) \circ \phi(s).$$

Therefore

$$\dot{\phi}(s) - \dot{\psi}(s) = \Big(X'(\Phi(s, t_0, x_0)) - X'(\Phi(s, t, x))\Big) \circ \phi(s)$$
$$+ X'(\Phi(s, t, x)) \circ (\phi(s) - \psi(s)).$$

This expression will enable us to majorize $\psi(s) - \phi(s)$.

Let $V = \bar{K}_{\frac{\epsilon}{2}} \times J_{\frac{\epsilon}{2}} \times B_r$. We set

$$M_1 = \sup_{(s,t,x)\in V} |X'(\Phi(s,t_0,x_0)) - X'(\Phi(s,t,x))|_{\mathcal{L}(\mathcal{E})}$$

and

$$M_2 = \sup_{(s,t,x)\in V} |X'(\Phi(s,t,x))|_{\mathcal{L}(\mathcal{E})}.$$

As X' and Φ are continuous, if ϵ and r are sufficiently small, then M_1 is close to 0 and M_2 close to $|X'(\Phi(s_0,t_0,x_0))|_{\mathcal{L}(\mathcal{E})}$. Using the fact that $\phi(t_0) = \psi(t) = \mathrm{id}_E$, we obtain

$$\phi(t) - \psi(t) = \int_{t_0}^{t} \dot{\phi}(u)\mathrm{d}u,$$

which implies that

$$|\phi(t) - \psi(t)|_{\mathcal{L}(E)} \le |t - t_0|\gamma,$$

where $\gamma = \sup_{u\in\bar{K}_{\frac{\epsilon}{2}}} |\dot{\phi}(u)|_{\mathcal{L}(E)}$. Therefore, for $s \in \bar{K}_{\frac{\epsilon}{2}}$ and $s \ge t$, we have

$$|\phi(s) - \psi(s)|_{\mathcal{L}(E)} \le |\phi(t) - \psi(t)|_{\mathcal{L}(E)} + M_1 \int_{t}^{s} |\phi(u)|_{\mathcal{L}(E)}\mathrm{d}u$$

$$+ \int_{t}^{s} M_2|\phi(u) - \psi(u)|_{\mathcal{L}(E)}\mathrm{d}u$$

$$\le |t - t_0|\gamma + M_1\delta + M_2 \int_{t}^{s} |\phi(u) - \psi(u)|_{\mathcal{L}(E)}\mathrm{d}u,$$

where $\delta = \int_{\bar{K}_{\frac{\epsilon}{2}}} |\phi(u)|_{\mathcal{L}(E)}\mathrm{d}u$. Setting $R = |t - t_0|\gamma + M_1\delta$ and applying Corollary 9.1, we obtain

$$|\phi(s) - \psi(s)|_{\mathcal{L}(E)} \le Re^{M_2(s-t)}.$$

If $s \in \bar{K}_{\frac{\epsilon}{2}}$ and $s < t$, then in a similar way we obtain

$$|\phi(s) - \psi(s)|_{\mathcal{L}(E)} \le |t - t_0|\gamma + M_1\delta + M_2 \int_{s}^{t} |\phi(u) - \psi(u)|_{\mathcal{L}(E)}\mathrm{d}u$$

and, using Exercise 9.3,

$$|\phi(s) - \psi(s)|_{\mathcal{L}(E)} \le Re^{M_2(t-s)}.$$

As R converges to 0, when ϵ and r approach 0, and

$$\phi(s_0) - \psi(s) = \phi(s_0) - \phi(s) + \phi(s) - \psi(s),$$

when (s, t, x) converges to (s_0, t_0, x_0), $\psi(s)$ converges $\phi(s_0)$, i.e., $\partial_3 \Phi(s, t, x)$ converges to $\partial_3 \Phi(s_0, t_0, x_0)$. Therefore $\partial \Phi_3$ is continuous at (s_0, t_0, x_0).

We may apply a similar argument to obtain the continuity at (s_0, t_0, x_0), where $s_0 < t_0$. This ends the proof.   □

Using the three propositions which we have just proved, we now have

Theorem 10.2. *If X is a vector field of class C^1 defined on an open subset of a Banach space E, then the flow Φ of X is also of class C^1.*

Proof. It is sufficient to apply Theorem 3.6.   □

In the next section we will generalize this result to the case where the vector field is of class C^k for any given $k \in \mathbb{N}^*$.

10.3   Higher Differentiability of the Flow

We have seen that, if a vector field X is of class C^1, then its flow is also of class C^1. We will now generalize this result. We will begin with a preliminary proposition. Suppose that E_1, E_2 and F are normed vector spaces, O an open subset of $E_1 \times E_2$ and $f : O \longrightarrow F$ a C^2-mapping. If $a \in O$, then $\partial_{ij} f(a) \in \mathcal{L}(E_i, \mathcal{L}(E_j, F))$. We may consider $\partial_{ij} f(a)$ to be a continuous bilinear mapping from $E_i \times E_j$ into F.

Proposition 10.4. *Suppose that E_1, E_2 and F are normed vector spaces, O an open subset of $E_1 \times E_2$ and $f : O \longrightarrow F$ a C^2-mapping. If $a \in O$, $x \in E_1$ and $v \in E_2$, then*

$$\partial_{12} f(a)(x, v) = \partial_{21} f(a)(v, x).$$

Proof. For $(u, v) \in E_1 \times E_2$ we have

$$f'(a)(u, v) = \partial_1 f(a)u + \partial_2 f(a)v$$

and so, with $(x, y) \in E_1 \times E_2$,

$$f^{(2)}(a)\big((x, y), (u, v)\big) = \partial_{11} f(a)(x, u) + \partial_{21} f(a)(y, u)$$
$$+ \partial_{12} f(a)(x, v) + \partial_{22} f(a)(y, v).$$

Changing the roles of (x, y) and (u, v), we have

$$f^{(2)}(a)\big((u, v), (x, y)\big) = \partial_{11} f(a)(u, x) + \partial_{21} f(a)(v, x)$$
$$+ \partial_{12} f(a)(u, y) + \partial_{22} f(a)(v, y).$$

As $f^{(2)}(a)$ is symmetric,

$$f^{(2)}(a)((x, y), (u, v)) = f^{(2)}(a)((u, v), (x, y)).$$

If we set $u = 0$ and $y = 0$, then we obtain

$$\partial_{12} f(a)(x, v) = \partial_{21} f(a)(v, x),$$

the desired result. $\square$

If $I \subset \mathbb{R}$ is an open interval, E a normed vector space, O an open subset of E and f a C^2-mapping from $I \times O$ into E, then both $\frac{d}{dt} \partial_2 f(t, x)$ and $\partial_2 \frac{d}{dt} f(t, x)$ belong to $\mathcal{L}(E)$.

Corollary 10.1. *We have*

$$\frac{d}{dt} \partial_2 f(t, x) = \partial_2 \frac{d}{dt} f(t, x)$$

for $t \in I$ *and* $x \in O$.

Proof. We claim that

$$\partial_{12} f(t, x)(s, v) = \left(s \frac{d}{dt} \partial_2 f(t, x) \right) v \text{ and } \partial_{21} f(t, x)(v, s) = s \left(\partial_2 \frac{d}{dt} f(t, x) v \right).$$

First we have

$$\partial_{12} f(t, x)(s, v) = (\partial_{12} f(t, x)s)v.$$

$\partial_2 f(t, x)$ is a mapping from $\mathbb{R} \times E$ into $\mathcal{L}(E)$. Then ∂_{12} is the differential of this mapping with respect to the first variable. We thus obtain

$$(\partial_{12} f(t, x)s)v = \left(s \frac{d}{dt} \partial_2 f(t, x) \right) v.$$

We now turn to the second part of the claim; to begin

$$\partial_{21} f(t, x)(v, s) = (\partial_{21} f(t, x)v)s.$$

$\partial_1 f(t, x)$ is a mapping from $\mathbb{R} \times E$ into $\mathcal{L}(\mathbb{R}, E)$. Then $\partial_{21} f(t, x)$ is the differential of this mapping with respect to the second variable. If α is the mapping from E into $\mathcal{L}(\mathbb{R}, E)$ defined by

$$\alpha(x) = (s \longmapsto sx),$$

then

$$\partial_1 f(t, x) = \alpha \circ \frac{d}{dt} f(t, x)$$

and so

$$\partial_{21} f(t, x)v = \alpha \left(\partial_2 \frac{d}{dt} f(t, x)v \right) \implies (\partial_{21} f(t, x)v)s = s \left(\partial_2 \frac{d}{dt} f(t, x)v \right).$$

This proves the second part of the statement we claimed.

Now, setting $s = 1$ and using the proposition, we obtain

$$\frac{\mathrm{d}}{\mathrm{d}t}\partial_2 f(t,x)v = \partial_{12}f(t,x)(1,v) = \partial_{21}f(t,x)(v,1) = \partial_2\frac{\mathrm{d}}{\mathrm{d}t}f(t,x)v$$

for all $v \in E$, which implies that

$$\frac{\mathrm{d}}{\mathrm{d}t}\partial_2 f(t,x) = \partial_2\frac{\mathrm{d}}{\mathrm{d}t}f(t,x).$$

This ends the proof. $\square$

We now prove the main result of this section, namely

Theorem 10.3. *Let X be a vector field defined on an open subset O of a Banach space E. If X is of class C^k with $k \geq 1$, then the flow Φ of X is of class C^k.*

Proof. We will prove the result by induction on k. For $k = 1$ we have already proved it. Suppose now that the result is true up to k and consider the case $k + 1$. From Proposition 4.7 the mapping

$$H : \mathcal{L}(\mathbb{R}, E) \times \mathcal{L}(\mathbb{R}, E) \times \mathcal{L}(E) \longrightarrow \mathcal{L}(\mathbb{R} \times \mathbb{R} \times E, E)$$

defined by

$$H(f,g,h)(u,v,w) = f(u) + g(v) + h(w)$$

is a normed vector space isomorphism. Clearly

$$\Phi'(s,t,x) = H\big(\partial_1\Phi(s,t,x), \partial_2\Phi(s,t,x), \partial_3\Phi(s,t,x)\big).$$

Hence, because H is smooth, to show that Φ is of class C^{k+1} it is sufficient to prove that the three partial differentials are of class C^k.

First notice that

$$\frac{\partial}{\partial s}\Phi(s,t,x) = X(\Phi(s,t,x)).$$

X is of class C^{k+1} and so *a fortiori* of class C^k. By hypothesis Φ is also of class C^k and hence $\frac{\partial}{\partial s}\Phi$ is of class C^k. However, the mapping from E into $\mathcal{L}(\mathbb{R}, E)$ defined by

$$\alpha(x) = (u \longmapsto ux)$$

is an isometric isomorphism and $\alpha \circ (\frac{\partial}{\partial s}\Phi) = \partial_1\Phi$. Therefore $\partial_1\Phi$ is of class C^k.

We have seen in Sect. 10.2 that

$$\frac{\partial}{\partial t}\Phi(s,t,x) = -X(\Phi(s,t,x))$$

and so $\frac{\partial}{\partial t}\Phi$ is of class C^k. As $\alpha(\frac{\partial}{\partial t}\Phi) = \partial_2\Phi$, $\partial_2\Phi$ is of class C^k.

We now consider the third partial differential. This is a little more difficult. We define a vector field Z on $O \times \mathcal{L}(E)$ by

$$Z(x, L) = (X(x), X'(x) \circ L).$$

As both coordinate mappings are of class C^k, so is Z. By hypothesis the flow of Z is of class C^k. For $(s, t, x) \in D$, the domain of Φ, and $L \in \mathcal{L}(E)$ let us set

$$\Psi(s, t, (x, L)) = (\Phi(s, t, x), \partial_3 \Phi(s, t, x) \circ L).$$

We claim that Ψ is the flow of Z. First we have

$$\Psi(t, t, (x, L)) = (\Phi(t, t, x), \partial_3 \Phi(t, t, x) \circ L) = (x, \mathrm{id}_E \circ L) = (x, L).$$

Now, using Corollary 10.1, we have

$$\frac{\partial}{\partial s} \partial_3 \Phi(s, t, x) \circ L = \partial_3 \frac{\partial}{\partial s} \Phi(s, t, x) \circ L$$

$$= \partial_3 X(\Phi(s, t, x)) \circ L$$

$$= X'(\Phi(s, t, x)) \circ \partial_3 \Phi(s, t, x) \circ L$$

and hence

$$\frac{\partial}{\partial s} \Psi(s, t, (x, L)) = Z(\Psi(s, t, (x, L))).$$

Therefore Ψ is the flow of Z. It follows that the second coordinate of Ψ is of class C^k. As the mapping $(s, t, x) \longmapsto (s, t, x, \mathrm{id}_E)$ is smooth, $\partial_3 \Phi$ is of class C^k. This finishes the proof.   □

10.4 The Reduced Flow

Let E be a Banach space, X a vector field defined on an open subset of E, with flow $\Phi(s, t, x)$. If we fix the second variable t, then we obtain a mapping from a subset of $\mathbb{R} \times E$ into E; in particular, if we take $t = 0$. The mapping

$$\phi(s, x) = \Phi(s, 0, x)$$

is called the *reduced flow* of X. (The reduced flow is often referred to as the flow; however, we will reserve this term for the mapping Φ already defined.) We will sometimes use the abbreviation *r-flow* for reduced flow.

Theorem 10.4. *Let X be a vector field defined on an open subset of a Banach space E. The r-flow ϕ of X is defined on an open subset Ω of $\mathbb{R} \times E$. If X is of class C^k with $k \geq 1$, then ϕ is also of class C^k.*

Proof. Let α be the mapping from $\mathbb{R} \times E$ into $\mathbb{R} \times \mathbb{R} \times E$ defined by

$$\alpha(s, x) = (s, 0, x).$$

α is a continuous linear mapping, hence smooth. If D is the domain of the flow Φ, then $\Omega = \alpha^{-1}(D)$. As D is open, so is Ω.

Suppose that X is of class C^k. As $\phi = \Phi \circ \alpha$ and Φ is of class C^k, ϕ is of class C^k.   □

Earlier we stated without proof that the exponential mapping from a Banach algebra into itself is smooth. We are now in a position to give a simple proof of this assertion.

Theorem 10.5. *If E is Banach algebra, then the exponential mapping defined on E is smooth.*

Proof. Let X be the vector field defined on E^2 by

$$X(x, y) = (yx, 0)$$

The mapping $(x, y) \longmapsto yx$ is bilinear and continuous and so smooth. As the mapping $(x, y) \longmapsto 0$ is constant and so smooth, X is smooth. Therefore the reduced flow ϕ of X is smooth. If 1_E is the identity for the multiplication in E, then we have

$$\phi(t, (1_E, y)) = (\exp(ty), y)$$

for $(t, y) \in \mathbb{R} \times E$. As the mapping $y \longmapsto (1_E, y)$ is smooth, so is the mapping

$$\psi : \mathbb{R} \times E \longrightarrow E \times E, (t, y) \longmapsto (\exp(ty), y).$$

The coordinate mappings of ψ are smooth, in particular the first one, i.e., the mapping $(t, y) \longmapsto \exp(ty)$. This in turn implies that

$$\exp(y) = \exp(1y)$$

for $y \in E$, is smooth.   □

Remark. As $\exp'(0) = \mathrm{id}_E$ and $\exp(0) = 1_E$, there are open neighbourhoods U and V of respectively of 0 and 1_E such that $\exp(U) = V$ and $\exp : U \longrightarrow V$ is a smooth diffeomorphism.

Let us now return to general vector fields. From the definition of the reduced flow ϕ we have

$$\phi(0, x) = x.$$

If we fix x and set

$$\phi_x(s) = \phi(s, x),$$

then the mapping ϕ_x is defined on an open interval I_x of $\mathbb{R}$ containing 0. As the mapping $\beta : s \longmapsto (s, x)$ is smooth, if ϕ is of class C^k, then so is ϕ_x. In the same way we may define a mapping ϕ_s by fixing s:

$$\phi_s(x) = \phi(s, x).$$

The mapping $\gamma : x \longmapsto (s, x)$ is smooth and the domain of ϕ_s is $\gamma^{-1}(\Omega)$. Therefore the domain of ϕ_s is open in E. For $s = 0$ the domain of ϕ_s is clearly that of the vector field. However, this is in general not the case and the domain of ϕ_s may even be empty for some values of s. If the domain of ϕ_s is not empty and ϕ is of class C^k, then so is ϕ_s.

Example. In Sect. 9.1 we saw that the maximal integral curve of the vector field

$$X : \mathbb{R} \longrightarrow \mathbb{R}, x \longmapsto 1 + x^2,$$

with initial conditions $(0, x_0)$, is defined on the interval $(-\frac{\pi}{2} - \arctan x_0, \frac{\pi}{2} + \arctan x_0)$. Therefore, if $s \notin (-\pi, \pi)$, then the domain of ϕ_s is empty.

If X is a perfect vector field and O its domain, then ϕ_x is defined on $\mathbb{R}$ for any $x \in O$, and ϕ_s is defined on O for any $s \in \mathbb{R}$.

In Sect. 9.1 we obtained the following result: if D is the domain of Φ and $(t_1, 0, x_0) \in D$, then $(t, 0, x_0) \in D$ if and only if $(t - t_1, 0, \Phi(t_1, 0, x_0)) \in D$ and in this case

$$\Phi(t, 0, x_0) = \Phi(t - t_1, 0, \Phi(t_1, 0, x_0)).$$

We may write this in the following form: if $u \in I_x$, then $s + u \in I_x$ if and only if $s \in I_{\phi_u(x)}$ and in this case

$$\phi_{s+u}(x) = \phi_s \circ \phi_u(x).$$

Suppose that the domain of ϕ_s is not empty. Then setting $u = -s$ we obtain the following result: the mapping ϕ_s is invertible and $\phi_s^{-1} = \phi_{-s}$. In addition, if X is of class C^k with $k \geq 1$, then ϕ_s is a C^k-diffeomorphism.

Exercise 10.1. We may define a relation R on the domain O of a vector field X by setting xRy when there is a $t \in \mathbb{R}$ with $\phi(t, x) = y$. Show that this relation is an equivalence relation. Thus the vector field X generates a partition of O.

10.5   One-Parameter Subgroups

If X is a complete vector field defined on an open subset O of a Banach space E, then for every $t \in \mathbb{R}$ the mapping ϕ_t is defined on O. Each ϕ_t is a diffeomorphism from O onto itself. The mappings $(\phi_t)_{t \in \mathbb{R}}$ form a subgroup of the group of diffeomorphisms from O into itself. We have

- $\phi_s \circ \phi_t = \phi_{s+t}$;
- $\phi_0 = \mathrm{id}_O$;
- $\phi_t^{-1} = \phi_{-t}$.

If X is of class C^k, then ϕ_t is of class C^k and in this case the mappings $(\phi_t)_{t \in \mathbb{R}}$ form a subgroup of the group of C^k-diffeomorphisms from O onto itself. The collection of mappings $(\phi_t)_{t \in \mathbb{R}}$ form what is called a one-parameter subgroup. Here is an formal definition. Suppose that O is an open subset of a Banach space E and that $(\psi_t)_{t \in \mathbb{R}}$ is a collection of diffeomorphisms from O onto itself. Then $(\psi_t)_{t \in \mathbb{R}}$ is a *one-parameter subgroup* if

- The mapping $\psi : \mathbb{R} \times O \longrightarrow O, (t, x) \longmapsto \psi_t(x)$ is of class C^1;
- $\psi_0 = \mathrm{id}_E$ and $\psi_s \circ \psi_t = \psi_{s+t}$ for $s, t \in \mathbb{R}$.

We say that the one-parameter subgroup is of class C^k if the mapping ψ is of class C^k. Clearly, if the vector field X is complete and of class C^k with $k \geq 1$, then the collection of mappings $(\phi_t)_{t \in \mathbb{R}}$ is a one-parameter subgroup of class C^k. In fact, as we will now see, all one-parameter subgroups are essentially of this form.

Proposition 10.5. *Suppose that $(\psi_t)_{t \in \mathbb{R}}$ is a one-parameter subgroup of diffeomorphisms defined on an open subset O of a Banach space E and that the mapping $\psi(t, x)$ is of class C^k with $k \geq 2$. If we set*

$$X(x) = \frac{\mathrm{d}}{\mathrm{d}t} \psi(t, x)_{|t=0}$$

for $x \in O$, then X is a complete vector field of class C^{k-1}, whose reduced flow is ψ.

Proof. Clearly X is a mapping from O into E. In addition,

$$\frac{\mathrm{d}}{\mathrm{d}t} \psi(t, x)_{|t=0} = \partial_1 \psi(0, x) 1.$$

From Theorem 4.7 $\partial_1 \psi$ is of class C^{k-1} and it follows that X is a vector field of class C^{k-1} (see exercise below).

Now,

$$X(\psi(s, x)) = \frac{\mathrm{d}}{\mathrm{d}t} \psi(t, \psi(s, x))_{|t=0} = \frac{\mathrm{d}}{\mathrm{d}t} \psi(t + s, x))_{|t=0} = \frac{\mathrm{d}}{\mathrm{d}s} \psi(s, x),$$

therefore $\psi(s, x)$ is an integral curve. This curve is defined on $\mathbb{R}$ and $\psi(0, x) = x$. Hence $\psi(s, x)$ is the maximal integral curve with initial conditions $(0, x)$. If α is the maximal integral curve with initial conditions (t, x), then $\alpha(s) = \psi(s - t, x)$ and so α is defined on $\mathbb{R}$. We have shown that X is complete and has ψ for reduced flow. $\square$

Exercise 10.2. Show that the mapping

$$O \longrightarrow E, x \longmapsto \partial_1 \psi(0, x) 1$$

is of class C^{k-1}.

Remark. The vector field X is called the *infinitesimal generator* of the one-parameter subgroup.

The next result follows immediately from the proposition above.

Theorem 10.6. *There is a 1–1 correspondence between smooth complete vector fields and smooth one-parameter subgroups.*

Chapter 11
The Calculus of Variations: An Introduction

If $\mathcal{E}$ is a normed vector space composed of mappings, for example, the space of continuous real-valued functions defined on a compact interval, then we refer to a real-valued mapping F defined on a subset S of $\mathcal{E}$ as a *functional* . The calculus of variations is concerned with the search for extrema of functionals. In general, the set S is determined, at least partially, by constraints on the mappings and the functional F is defined by an integral. The elements of S are often said to be F-*admissible* (or admissible if there is no possible confusion).

Many physical problems involve looking for best trajectories in some sense. For example, one might consider crossing a river from one point to another in the quickest possible time or finding the shortest way from going from one point to another on some surface. Such problems lead in general to the minimization of the integral of a mapping defined on a compact interval with given values for the endpoints. We will concentrate on this classical problem, called a *fixed endpoint problem*. Unlike for optimization problems in finite-dimensional spaces, critical points (and hence extrema) in infinite-dimensional spaces are usually difficult to determine. However, the critical points of fixed endpoint problems can in general be characterized as solutions of differential equations, which can often be solved. This makes fixed endpoint problems particularly interesting.

11.1 The Space $C^1(I, E)$

In this section we introduce a normed vector space which is important both in the calculus of variations and elsewhere. Let E be a normed vector space and $I = [a, b]$ a compact interval of $\mathbb{R}$. The space of continuous mappings γ from I into E, which we note $C^0(I, E)$ (or $C(I, E)$), is a vector space. This space has the natural norm:

$$\|\gamma\|_{C^0} = \sup_{t \in I} \|\gamma(t)\|_E.$$

R. Coleman, *Calculus on Normed Vector Spaces*, Universitext,
DOI 10.1007/978-1-4614-3894-6_11, © Springer Science+Business Media New York 2012

If E is a Banach space, then so is $C^0(I, E)$ (see Exercise 1.23). We recall that a mapping γ from I into E is of class C^1 if it has derivatives at all points $t \in I$ and the mapping $\dot{\gamma}$ is continuous. The set of all C^1-mappings (or curves) γ defined on I form a vector space which we note $C^1(I, E)$. This space has a natural norm:

$$\|\gamma\|_{C^1} = \|\gamma\|_{C^0} + \|\dot{\gamma}\|_{C^0}.$$

Proposition 11.1. *If E is a Banach space and I is a nontrivial closed interval $[a, b]$, then $C^1(I, E)$ is a Banach space.*

Proof. Let (γ_n) be a Cauchy sequence in $C^1(I, E)$. Then (γ_n) and $(\dot{\gamma}_n)$ are Cauchy sequences in $C^0(I, E)$, hence (γ_n) has a limit $\gamma \in C^0(I, E)$ and $(\dot{\gamma}_n)$ has a limit $\delta \in C^0(I, E)$. As $\dot{\gamma}_n$ is continuous and γ_n a primitive of $\dot{\gamma}_n$, we have

$$\gamma_n(x) - \gamma_n(a) = \int_a^x \dot{\gamma}_n(t)\mathrm{d}t$$

for $x \in I$. Also,

$$\|\gamma(x) - \gamma(a) - \int_a^x \delta(t)\mathrm{d}t\|_E \leq \|\gamma(x) - \gamma_n(x)\|_E + \|\gamma_n(a) - \gamma(a)\|_E$$

$$+ \|\int_a^x \dot{\gamma}_n(t) - \delta(t)\mathrm{d}t\|_E$$

$$\leq 2\|\gamma - \gamma_n\|_{C^0} + (b - a)\|\dot{\gamma}_n - \delta\|_{C^0}.$$

As this last expression converges to 0 when n goes to ∞, we obtain

$$\gamma(x) = \gamma(a) + \int_a^x \delta(t)\mathrm{d}t.$$

It follows that $\dot{\gamma} = \delta$ and so the sequence (γ_n) converges to γ in $C^1(I, E)$.   □

The normed vector space of mappings for the fixed endpoint problem will be $C^1(I, E)$, for some compact interval I and Banach space E.

11.2   Lagrangian Mappings

As in the previous section, we suppose that $I = [a, b]$ is a compact interval of $\mathbb{R}$ and E a Banach space. The functional $\mathcal{L}$ which we will define on a subset of $\mathcal{E} = C^1(I, E)$ will be the integral of an expression involving a curve in $\mathcal{E}$ and its derivative. However, before defining the functional we will do some preliminary work.

Let O is an open subset of $\mathbb{R} \times E \times E$. A continuous real-valued function L defined on O is called a *Lagrangian function* (or Lagrangian). A curve $\gamma \in C^1(I, E)$ is said to be *L-admissible* (or admissible if there is no possible confusion) if $(t, \gamma(t), \dot{\gamma}(t))$ belongs to O for all $t \in I$.

Proposition 11.2. *The set of L-admissible curves, which we note Ω_L (or Ω), is open in $C^1(I, E)$.*

Proof. If Ω is empty, then Ω is open. Suppose that this is not the case and consider the mapping

$$\psi : I \times C^1(I, E) \longrightarrow \mathbb{R} \times E \times E, (t, \gamma) \longmapsto (t, \gamma(t), \dot{\gamma}(t)).$$

The coordinate mappings of ψ are clearly continuous, hence ψ is continuous. Let us fix $\gamma \in \Omega$ and take $t \in I$. As ψ is continuous, there is an open interval J_t of $\mathbb{R}$ containing t, and an open ball $B(\gamma, r_t) \subset C^1(I, E)$ such that $(s, \delta(s), \dot{\delta}(s)) \in O$ when $s \in J_t \cap I$ and $\delta \in B(\gamma, r_t)$. As I is compact, a finite number of the intervals J_t cover I. Let us note these intervals $J_{t_1}, \ldots, J_{t_p}$ and set $r = \min(r_{t_1}, \ldots, r_{t_p})$. If $s \in I$ and $\delta \in B(\gamma, r)$, then $(s, \delta(s), \dot{\delta}(s)) \in O$ and hence δ is admissible. Therefore the open ball $B(\gamma, r)$ lies in Ω and it follows that Ω is open. □

Examples. 1. If $E = \mathbb{R}$, $O = \mathbb{R}^3$, $I = [a, b]$ and L_1 is the real-valued function defined by

$$L_1(t, x, y) = e^t x - 3y^4,$$

then $\Omega_{L_1} = C^1(I, \mathbb{R})$.

2. If $E = \mathbb{R}$, $O = \mathbb{R} \times \mathbb{R} \times (-1, 1)$, $I = [a, b]$ and L_2 is the real-valued function defined by

$$L_2(t, x, y) = \frac{x}{\sqrt{1 - y^2}},$$

then

$$\Omega_{L_2} = \{\gamma \in C^1(I, \mathbb{R}) : \dot{\gamma}(t) \in (-1, 1), \ t \in I\}.$$

If $\gamma \in \Omega_L$, then the function

$$t \longmapsto L(t, \gamma(t), \dot{\gamma}(t))$$

is defined and continuous on $I = [a, b]$. We define a mapping $\mathcal{L}$ from Ω_L into $\mathbb{R}$ by

$$\mathcal{L}(\gamma) = \int_a^b L(t, \gamma(t), \dot{\gamma}(t)) dt.$$

$\mathcal{L}$ is called the *Lagrangian functional* associated with L. Notice that here the $\mathcal{L}$-admissible curves are the L-admissible curves.

Suppose now that L is of class C^1. We will consider the differentiability of $\mathcal{L}$.

Proposition 11.3. *If L is of class C^1, then $\mathcal{L}$ is of class C^1.*

Proof. If f is the real-valued function defined on $I \times \Omega_L$ by

$$f(t, \gamma) = L(t, \gamma(t), \dot{\gamma}(t)),$$

then $f = L \circ \psi$, where ψ is the mapping defined in the proof of Proposition 11.2. As L and ψ are continuous, f is continuous.

We now consider the partial differential $\partial_2 \psi$. We have

$$\psi(t, \gamma + h) - \psi(t, \gamma) = (0, h(t), \dot{h}(t)).$$

The mapping

$$\eta : C^1(I, E) \longmapsto I \times E \times E, h \longmapsto (0, h(t), \dot{h}(t))$$

is linear and

$$\|\eta(h)\|_{I \times E \times E} = \max(|0|, \|h(t)\|_E, \|\dot{h}(t)\|_E) \leq \|h\|_{C^1},$$

therefore η is continuous. It follows that

$$\partial_2 \psi(t, \gamma) h = (0, h(t), \dot{h}(t)).$$

Thus $\partial_2 \psi$ exists at any point (t, γ). Being constant, it is also continuous.

Now let us look at f. Applying the chain rule to $f = L \circ \psi$ with t fixed, we obtain

$$\partial_2 f(t, \gamma) = L'(\psi(t, \gamma)) \circ \partial_2 \psi(t, \gamma).$$

As $\partial_2 \psi$ is continuous on $I \times \Omega_L$, $\partial_2 f$ is also continuous on $I \times \Omega_L$. From Theorem 3.10 $\mathcal{L}$ is of class C^1 and

$$\mathcal{L}'(\gamma) = \int_a^b \partial_2 f(t, \gamma) \mathrm{d}t$$

or, using Proposition 3.7,

$$\mathcal{L}'(\gamma) h = \int_a^b \partial_2 f(t, \gamma) h \mathrm{d}t$$

for $h \in C^1(I, E)$. $\qquad\qquad\qquad\qquad\qquad\qquad\qquad\qquad\qquad\qquad\qquad\square$

Remark. We can say a little more:

$$\begin{aligned}
\partial_2 f(t, \gamma) h &= L'(\psi(t, \gamma)) \circ \partial_2 \psi(t, \gamma) h \\
&= L'(\psi(t, \gamma))(0, h(t), \dot{h}(t)) \\
&= \partial_2 L(\psi(t, \gamma)) h(t) + \partial_3 L(\psi(t, \gamma)) \dot{h}(t)
\end{aligned}$$

and so

$$\mathcal{L}'(\gamma)h = \int_a^b \partial_2 f(t,\gamma)h dt = \int_a^b \Big(\partial_2 L(\psi(t,\gamma))h(t) + \partial_3 L(\psi(t,\gamma))\dot{h}(t)\Big)dt.$$

From here on we will suppose that the Lagrangian L is of class C^1.

11.3   Fixed Endpoint Problems

Suppose now that we aim to minimize $\mathcal{L}$ and that at the same time we require that the endpoints of the curves $\gamma \in \Omega_L$ have certain values: $\gamma(a) = \alpha$, $\gamma(b) = \beta$. In this case we consider $\mathcal{L}$ restricted to the space

$$\Omega_L(\alpha, \beta) = \{\gamma \in \Omega_L : \gamma(a) = \alpha, \gamma(b) = \beta\},$$

i.e., we search for mimima of the functional $\mathcal{L}$ on the set $S = \Omega_L(\alpha, \beta)$.

If we set

$$A(\alpha, \beta) = \{\gamma \in C^1(I, E) : \gamma(a) = \alpha, \gamma(b) = \beta\},$$

then $A(\alpha, \beta)$ is a closed affine subspace of $C^1(I, E)$. It is the translation of the closed vector space $\mathcal{A}(0, 0)$ by the curve δ defined by

$$\delta((1 - s)a + sb) = (1 - s)\alpha + s\beta$$

with $s \in [0, 1]$. To simplify the notation we will write $\mathcal{A}$ for $\mathcal{A}(0, 0)$. Clearly

$$\Omega_L(\alpha, \beta) = \Omega_L \cap A(\alpha, \beta).$$

From Corollary 2.4, if γ is a minimum, then

$$\mathcal{L}'(\gamma)h = 0$$

for $h \in \mathcal{A}$. A curve $\gamma \in C^1(I, E)$ satisfying this condition is called an *extremal* of the Lagrangian functional $\mathcal{L}$ with fixed endpoints a and b. It should be noticed that an extremal is not necessarily an extremum (maximum or minimum) of the fixed endpoint problem under consideration. An extremum is an extremal, but an extremal is not necessarily an extremum. (It would probably be better to use the term critical curve instead of extremal.) In the next section we will see that this condition leads us to a differential equation called the Euler–Lagrange equation.

In the appendix to Chap. 9 we showed that, for any bounded open interval I of $\mathbb{R}$, there exists a smooth real-valued function g defined on $\mathbb{R}$, which is strictly positive on I and vanishes elsewhere. We will use this in the proof of the next result.

Proposition 11.4. *Let $I = [a, b]$ be a closed bounded interval of $\mathbb{R}$ and $A : I \longrightarrow E^*$ a continuous mapping such that*

$$\int_a^b A(t)v(t)\mathrm{d}t = 0$$

for every continuous function $v : I \longrightarrow E$ such that $\int_a^b v(t)\mathrm{d}t = 0$. Then A is constant.

Proof. We will argue by contradiction. If A is not constant, then there exist $t_1, t_2 \in \mathbb{R}$ such that

$$a < t_1 < t_2 < b \qquad \text{and} \qquad A(t_1) \neq A(t_2).$$

Without loss of generality we can suppose that there is an element $u \in E$ such that $A(t_1)u > A(t_2)u$.

We take $\alpha_1, \alpha_2 \in \mathbb{R}$ such that

$$A(t_1)u > \alpha_1 > \alpha_2 > A(t_2)u.$$

As A is continuous, there is an $\epsilon > 0$ such that

$$A(t)u > \alpha_1 \text{ for } t \in (t_1 - \epsilon, t_1 + \epsilon) \quad \text{and} \quad A(t)u < \alpha_2 \text{ for } t \in (t_2 - \epsilon, t_2 + \epsilon).$$

We can take ϵ sufficiently small so that

$$a < t_1 - \epsilon < t_1 + \epsilon < t_2 - \epsilon < t_2 + \epsilon < b.$$

Let $\lambda : \mathbb{R} \longrightarrow \mathbb{R}_+$ be a smooth real-valued function which is strictly positive on the interval $(-\epsilon, +\epsilon)$ and vanishes elsewhere. We define a function μ on $\mathbb{R}$ by

$$\mu(t) = \lambda(t - t_1) - \lambda(t - t_2).$$

Then μ is smooth and

$$\int_a^b \mu(t)\mathrm{d}t = \int_a^b \lambda(t - t_1)\mathrm{d}t - \int_a^b \lambda(t - t_2)\mathrm{d}t$$

$$= \int_{a-t_1}^{b-t_1} \lambda(s)\mathrm{d}s - \int_{a-t_2}^{b-t_2} \lambda(s)\mathrm{d}s = 0,$$

because $a - t_i < -\epsilon$ and $\epsilon < b - t_i$ for $i = 1, 2$. In addition,

$$t \in (t_1 - \epsilon, t_1 + \epsilon) \Longrightarrow t - t_2 < -\epsilon \quad \text{and} \quad t \in (t_2 - \epsilon, t_2 + \epsilon) \Longrightarrow t - t_1 > \epsilon$$

and so we have

$$\begin{cases} \mu(t) > 0 & \text{for } t \in (t_1 - \epsilon, t_1 + \epsilon) \\ \mu(t) < 0 & \text{for } t \in (t_2 - \epsilon, t_2 + \epsilon) \, . \\ \mu(t) = 0 & \text{otherwise} \end{cases}$$

We now define $v : [a, b] \longrightarrow E$ by $v(t) = \mu(t)u$. Then v is continuous and $\int_a^b v(t)\mathrm{d}t = 0$. Also,

$$\int_a^b A(t)v(t)\mathrm{d}t = \int_a^b \lambda(t - t_1)A(t)u\mathrm{d}t - \int_a^b \lambda(t - t_2)A(t)u\mathrm{d}t$$

$$= \int_{t_1-\epsilon}^{t_1+\epsilon} \lambda(t - t_1)A(t)u\mathrm{d}t - \int_{t_2-\epsilon}^{t_2+\epsilon} \lambda(t - t_2)A(t)u\mathrm{d}t$$

$$> \alpha_1 \int_{t_1-\epsilon}^{t_1+\epsilon} \lambda(t - t_1)\mathrm{d}t - \alpha_2 \int_{t_2-\epsilon}^{t_2+\epsilon} \lambda(t - t_2)\mathrm{d}t$$

$$= (\alpha_1 - \alpha_2) \int_{-\epsilon}^{+\epsilon} \lambda(s)\mathrm{d}s > 0.$$

This is a contradiction and it follows that A is constant.   □

Example. Let L be the Lagrangian defined on $\mathbb{R}^3$ by

$$L(t, x, y) = y^2$$

and $I = [a, b]$, a closed bounded interval of $\mathbb{R}$. We aim to find a curve $\gamma \in C^1(I, R)$, with $\gamma(a) = 0$ and $\gamma(b) = 1$, such that γ minimizes the Lagrangian functional

$$\mathcal{L}(\gamma) = \int_a^b L(t, \gamma(t), \dot{\gamma}(t))\mathrm{d}t = \int_a^b \dot{\gamma}^2(t)\mathrm{d}t.$$

If γ is a minimum, then

$$\mathcal{L}'(\gamma)h = \int_a^b 2\dot{\gamma}(t)\dot{h}(t)\mathrm{d}t = 0$$

for all $h \in C^1(I, \mathbb{R})$ such that $h(a) = h(b) = 0$. It follows that

$$\int_a^b 2\dot{\gamma}(t)v(t)\mathrm{d}t = 0$$

for every continuous function v on $[a, b]$ with $\int_a^b v(t)\mathrm{d}t = 0$. From Proposition 11.4 there is a constant c such that for all $t \in I$

$$\dot{\gamma}(t) = c.$$

It easily follows that the only possibility is the curve γ defined on I by

$$\gamma(t) = \frac{t-a}{b-a}.$$

As yet we do not know whether γ is a minimum. Assume that $\delta \in C^1(I, \mathbb{R})$, with $\delta(a) = 0$ and $\delta(b) = 1$. Then

$$\mathcal{L}(\gamma + \delta) - \mathcal{L}(\gamma) = \int_a^b (\dot{\gamma}(t) + \dot{\delta}(t))^2 - \dot{\gamma}(t)^2 \mathrm{d}t$$

$$= \int_a^b \dot{\delta}(t)^2 \mathrm{d}t + 2 \int_a^b \dot{\gamma}(t)\dot{\delta}(t)\mathrm{d}t$$

$$\geq 2 \int_a^b \dot{\gamma}(t)\dot{\delta}(t)\mathrm{d}t$$

$$= \frac{2}{b-a} \int_a^b \dot{\delta}(t)\mathrm{d}t = \frac{2}{b-a} > 0.$$

Therefore γ is the unique minimum of the problem.

11.4   Euler–Lagrange Equations

As mentioned in the last section, we can associate a differential equation to a fixed endpoint problem. However, we will need some preliminary results.

Proposition 11.5. *Let I be an interval of $\mathbb{R}$ and $A : I \longrightarrow E^*$ and $u : I \longrightarrow E$ both differentiable at $t \in I$. Then $Au : I \longrightarrow \mathbb{R}$ is differentiable at t and*

$$\frac{\mathrm{d}}{\mathrm{d}t}(A(t)u(t)) = \dot{A}(t)u(t) + A(t)\dot{u}(t).$$

Proof. For s nonzero such that $t + s \in I$ we have

$$\frac{1}{s}[A(t+s)u(t+s) - A(t)u(t)] - \dot{A}(t)u(t) - A(t)\dot{u}(t)$$

$$= \left(\frac{1}{s}[A(t+s)u(t+s) - A(t)u(t+s)] - \dot{A}(t)u(t+s) \right)$$

$$+ \left(\dot{A}(t)u(t+s) - \dot{A}(t)u(t) \right) + \left(\frac{1}{s}[A(t)u(t+s) - A(t)u(t)] - A(t)\dot{u}(t) \right).$$

Each of the three parts of the expression on the right-hand side of the equation converges to 0 when s approaches 0, hence the result.   □

We can now establish an important result which will enable us to characterize extremals by means of differential equations.

Theorem 11.1. *Let $A, B : I \longrightarrow E^*$ be continuous. Then*

$$\int_a^b A(t)\gamma(t) + B(t)\dot{\gamma}(t)\mathrm{d}t = 0$$

for all $\gamma \in C^1(I, E)$ such that $\gamma(a) = \gamma(b) = 0$ if and only if B is differentiable and $\dot{B} = A$.

Proof. Let A and B be continuous mappings from the interval $I = [a, b]$ into E^*. Suppose that B is differentiable and $A = \dot{B}$. If $\gamma \in C^1(I, E)$ and $\gamma(a) = \gamma(b) = 0$, then using Proposition 11.5 we have

$$\int_a^b A(t)\gamma(t) + B(t)\dot{\gamma}(t)\mathrm{d}t = \int_a^b \dot{B}(t)\gamma(t) + B(t)\dot{\gamma}(t)\mathrm{d}t$$

$$= \int_a^b \frac{\mathrm{d}}{\mathrm{d}t}(B(t)\gamma(t))\mathrm{d}t$$

$$= B(b)\gamma(b) - B(a)\gamma(a) = 0.$$

Suppose now that for any $\gamma \in C^1(I, E)$ such that $\gamma(a) = \gamma(b) = 0$ we have

$$\int_a^b A(t)\gamma(t) + B(t)\dot{\gamma}(t)\mathrm{d}t = 0.$$

If we set $A_1(t) = \int_a^t A(s)\mathrm{d}s$, then

$$\frac{\mathrm{d}}{\mathrm{d}t}(A_1(t)\gamma(t)) = A(t)\gamma(t) + A_1(t)\dot{\gamma}(t)$$

and so

$$\int_a^b A(t)\gamma(t) + B(t)\dot{\gamma}(t)\mathrm{d}t = \int_a^b \frac{\mathrm{d}}{\mathrm{d}t}(A_1(t)\gamma(t)) - A_1(t)\dot{\gamma}(t) + B(t)\dot{\gamma}(t)\mathrm{d}t$$

$$= [A_1(t)\gamma(t)]_a^b - \int_a^b ((A_1(t) - B(t))\dot{\gamma}(t)\mathrm{d}t$$

$$= -\int_a^b (A_1(t) - B(t))\dot{\gamma}(t)\mathrm{d}t.$$

It follows that

$$\int_a^b (A_1(t) - B(t))v(t)\mathrm{d}t = 0$$

for every continuous function $v : I \longrightarrow E$ such that $\int_a^b v(t)\mathrm{d}t = 0$. If we apply Proposition 11.4, then we see that $A_1 - B$ is constant and therefore B is differentiable and $\dot{B} = A$.   □

Using results from this and previous sections we now obtain a differential equation characterizing an extremal, called the *Euler–Lagrange equation*:

Theorem 11.2. *The curve* $\gamma \in C^1(I, E)$ *is an extremal of the fixed endpoint problem if and only if the mapping* $\partial_3 L(t, \gamma(t), \dot{\gamma}(t))$ *is differentiable and*

$$\frac{\mathrm{d}}{\mathrm{d}t}\partial_3 L(t, \gamma(t), \dot{\gamma}(t)) = \partial_2 L(t, \gamma(t), \dot{\gamma}(t)).$$

Proof. In Sect. 11.2 we obtained the following expression for the differential of the Lagrangian functional:

$$\mathcal{L}'(\gamma)h = \int_a^b \Big(\partial_2 L(\psi(t, \gamma))h(t) + \partial_3 L(\psi(t, \gamma))\dot{h}(t)\Big)\mathrm{d}t.$$

However, γ is an extremal if and only if

$$\mathcal{L}'(\gamma)h = 0$$

for all $h \in C^1([a, b])$ such that $h(a) = h(b) = 0$. If we now apply Theorem 11.1, with

$$A(t) = \partial_2 L(\psi(t, \gamma)) \qquad \text{and} \qquad B(t) = \partial_3 L(\psi(t, \gamma)),$$

then we obtain the result.   □

Let us look in more detail at the case where $E = \mathbb{R}^n$. Here the Lagrangian is defined on an open subset $O \subset \mathbb{R}^{2n+1}$ and the image of a curve γ lies in $\mathbb{R}^n$. With $x, y, h \in \mathbb{R}^n$ we have

$$\partial_2 L(t, x, y)h = \frac{\partial L}{\partial x_1}(t, x, y)h_1 + \cdots + \frac{\partial L}{\partial x_n}(t, x, y)h_n,$$

and

$$\partial_3 L(t, x, y)h = \frac{\partial L}{\partial y_1}(t, x, y)h_1 + \cdots + \frac{\partial L}{\partial y_n}(t, x, y)h_n$$

and the Euler–Lagrange equation becomes

$$\frac{\mathrm{d}}{\mathrm{d}t}\left(\frac{\partial L}{\partial y_1}(\psi(t, \gamma))\right)h_1 + \cdots + \frac{\mathrm{d}}{\mathrm{d}t}\left(\frac{\partial L}{\partial y_n}(\psi(t, \gamma))\right)h_n = \frac{\partial L}{\partial x_1}(\psi(t, \gamma))h_1 + \cdots$$

$$+ \frac{\partial L}{\partial x_n}(\psi(t, \gamma))h_n.$$

We thus obtain the equations

$$\frac{\mathrm{d}}{\mathrm{d}t}\left(\frac{\partial L}{\partial y_i}(t, \gamma(t), \dot\gamma(t))\right) = \frac{\partial L}{\partial x_i}(t, \gamma(t), \dot\gamma(t))$$

for $i = 1, \ldots, n$.

Example. Let L be the Lagrangian defined on $\mathbb{R}^3$ by

$$L(t, x, y) = 2tx - x^2 + y^2.$$

and $I = [0, 1]$. We consider the problem of finding a curve $\gamma \in C^1(I, R)$, with $\gamma(0) = 0$ and $\gamma(1) = 2$, which minimizes the Lagrangian functional

$$\mathcal{L}(\gamma) = \int_0^1 L(t, \gamma(t), \dot\gamma(t))\mathrm{d}t = \int_0^1 2t\gamma(t) - \gamma^2(t) + \dot\gamma^2(t)\mathrm{d}t.$$

We have

$$\frac{\partial L}{\partial x} = 2(t - x) \qquad \text{and} \qquad \frac{\partial L}{\partial y} = 2y$$

and the Euler–Lagrange equation is

$$2\frac{\mathrm{d}}{\mathrm{d}t}\dot\gamma(t) = 2(t - \gamma(t))$$

or

$$\ddot\gamma(t) + \gamma(t) = t.$$

Therefore γ is a solution of the differential equation

$$\ddot y + y = t,$$

whose general solution is

$$y(t) = \alpha \cos t + \beta \sin t + t,$$

with $\alpha, \beta \in \mathbb{R}$. Taking into account the endpoint values we obtain

$$\gamma(t) = \frac{\sin t}{\sin 1} + t,$$

which is the unique extremal. However, without further work, we do not know whether γ is a minimum or not.

To close this section, it should be observed that the Euler–Lagrange equation is in general not easy, if not impossible, to solve explicitly. However, even in this case, it may give us important information about extremals.

11.5   Convexity

If the Lagrangian functional $\mathcal{L}$ is g-convex on the set $\Omega_L(\alpha, \beta)$, then an extremal lying in $\Omega_L(\alpha, \beta)$ is a global minimum. This follows from Proposition 7.5. Therefore it is important to know whether $\mathcal{L}$ is g-convex. The next result gives a criterion for the g-convexity of the Lagrangian functional.

Theorem 11.3. *If the Lagrangian L satisfies the property*

$$L(t, x + h, y + k) - L(t, x, y) \geq \partial_2 L(t, x, y)h + \partial_3 L(t, x, y)k,$$

then the associated Lagrangian functional $\mathcal{L}$ is g-convex on $\Omega_L(\alpha, \beta)$.

Proof. If $\gamma, \gamma + \delta \in \Omega_L(\alpha, \beta)$, then

$$\mathcal{L}(\gamma + \delta) - \mathcal{L}(\gamma) = \int_a^b L(t, \gamma(t) + \delta(t), \dot{\gamma}(t) + \dot{\delta}(t))dt - \int_a^b L(t, \gamma(t), \dot{\gamma}(t))dt$$

$$\geq \int_a^b \partial_2 L(t, \gamma, \dot{\gamma}(t))\delta(t) + \partial_3 L(t, \gamma, \dot{\gamma}(t))\dot{\delta}(t)dt$$

$$= \mathcal{L}'(\gamma)\delta.$$

Therefore $\mathcal{L}$ is g-convex on $\Omega_L(\alpha, \beta)$.   □

We will refer to the condition of the theorem as condition (C)

Remark. Clearly, if the Lagrangian L is g-convex, then the condition (C) is satisfied; however, the converse is not true. For example, if

$$L(t, x, y) = t^3 + x + y^2$$

for $(t, x, y) \in \mathbb{R}^3$, then L satisfies condition (C), but L is not g-convex on $\mathbb{R}^3$.

Example. Let L be the Lagrangian defined on $\mathbb{R}^3$ by

$$L(t, x, y) = 2tx + x^2 + y^2$$

and $I = [0, 1]$. We consider the problem of finding a curve $\gamma \in C^1(I, R)$, with $\gamma(0) = 0$ and $\gamma(1) = 1$, which minimizes the Lagrangian functional

$$\mathcal{L}(\gamma) = \int_0^1 L(t, \gamma(t), \dot{\gamma}(t))dt = \int_0^1 2t\gamma(t) + \gamma^2(t) + \dot{\gamma}^2(t)dt.$$

We first look for an extremal. We have

$$\frac{\partial L}{\partial x} = 2(t + x) \qquad \text{and} \qquad \frac{\partial L}{\partial y} = 2y$$

and the Euler–Lagrange equation is

$$2\frac{d}{dt}\dot{\gamma}(t) = 2(t + \gamma(t)),$$

or

$$\ddot{\gamma}(t) - \gamma(t) = t.$$

Therefore γ is a solution of the differential equation

$$\ddot{y} - y = t,$$

whose general solution is

$$y(t) = \alpha e^t + \beta e^{-t} - t,$$

with $\alpha, \beta \in \mathbb{R}$. Taking into account the endpoint values we obtain

$$\gamma(t) = \frac{2 \sinh t}{\sinh(1)} - t,$$

which is the unique extremal. However, L satisfies the condition (C) and so the Lagrangian functional $\mathcal{L}$ is g-convex. It follows from Proposition 7.5 that γ is a global minimum (in fact, the unique global minimum).

Exercise 11.1. Let $L(t, x, y) = x + y^2$ and $I = [0, 1]$. Find the unique curve $\gamma \in C^1(I, \mathbb{R})$, with $\gamma(0) = 0$ and $\gamma(1) = 1$, which minimizes the Lagrangian functional

$$\mathcal{L}(\gamma) = \int_0^1 L(t, \gamma(t), \dot{\gamma}(t))dt.$$

11.6   The Class of an Extremal

Extremals are by definition of class C^1; however, they may be of a higher class. Suppose that the Lagrangian is of class C^2. The partial differential $\partial_3 L(t, x, y)$ belongs to E^* and so $\partial_{33} L(t, x, y)$ is a continuous linear mapping from E into E^*. If $\partial_{33} L(t, x, y)$ is a normed vector space isomorphism for all (t, x, y) in the domain of L, then we say that L is *regular*.

Theorem 11.4. *If the Lagrangian L is regular and of class C^k, with $k \geq 2$, on its domain O, then any extremal is of class C^k on the interior of its interval of definition.*

Proof. Let γ be an extremal defined on an interval I and $t_0 \in \text{int } I$. Let us set $x_0 = \gamma(t_0)$, $y_0 = \dot{\gamma}(t_0)$ and $z_0 = \partial_3 L(t_0, x_0, y_0)$. We consider the mapping

$$F : O \longrightarrow \mathbb{R} \times E \times E^*, (t, x, y) \longmapsto (t, x, \partial_3 L(t, x, y)).$$

F is differentiable and

$$F'(t_0, x_0, y_0)(s, h, k) = \Big(s, h, \partial_{13} L(t_0, x_0, y_0)s + \partial_{23} L(t_0, x_0, y_0)h$$

$$+ \partial_{33} L(t_0, x_0, y_0)k\Big).$$

As $\partial_{33}(t_0, x_0, y_0)$ is a normed vector space isomorphism from E onto E^*, $F'(t_0, x_0, y_0)$ is a normed vector space isomorphism from $\mathbb{R} \times E \times E$ onto $\mathbb{R} \times E \times E^*$. Also, F is of class C^{k-1}. From the inverse mapping theorem we know that there is a neighbourhood U of (t_0, x_0, y_0) and a neighbourhood V of (t_0, x_0, z_0) such that $F : U \longrightarrow V$ is a C^{k-1}-diffeomorphism. We can write

$$F^{-1}(t, x, z) = (t, x, h(t, x, z)),$$

where h is a mapping of class C^{k-1}.

Now let J be an open interval of $\mathbb{R}$ containing t_0 and W an open neighbourhood of (x_0, z_0) such that $J \times W \subset V$. We define a time-dependent vector field $Y : J \times W \longrightarrow E \times E^*$ by

$$Y(t, x, z) = \big(h(t, x, z), \partial_2 L(t, x, h(t, x, z))\big).$$

Y is of class C^{k-1} and so there is a maximal integral curve $\phi(t) = (x(t), z(t))$ such that $\phi(t_0) = (x_0, z_0)$ defined on an open interval $I \subset J$. This integral curve is of class C^{k-1}. In addition, $\dot{x}(t) = h(t, x(t), z(t))$ and so $\dot{x}(t)$ is of class C^{k-1}. It follows that $x(t)$ is of class C^k.

Let us now set

$$\rho(t) = \big(\gamma(t), \partial_3 L(t, \gamma(t), \dot{\gamma}(t))\big).$$

We aim to show that ρ is an integral curve of the time-dependent vector field Y. For t close to t_0 we have

$$h\big(t, \gamma(t), \partial_3 L(t, \gamma(t), \dot{\gamma}(t))\big) = \dot{\gamma}(t)$$

and, using the Euler–Lagrange equation,

$$\partial_2 L\Big(t, \gamma(t), h\big(t, \gamma(t), \partial_3 L(t, \gamma(t), \dot{\gamma}(t))\big)\Big) = \partial_2 L(t, \gamma(t), \dot{\gamma}(t))$$

$$= \frac{\mathrm{d}}{\mathrm{d}t} \partial_3 L(t, \gamma(t), \dot{\gamma}(t)).$$

It follows that $\rho(t)$ is an integral curve of Y.

To conclude, we notice that $\rho(t_0) = (x_0, z_0)$ and so $\rho(t) = \phi(t)$ on a neighbourhood of t_0. Therefore $\gamma(t) = x(t)$ and so γ is of class C^k on a neighbourhood of t_0. This finishes the proof. $\qquad \square$

Suppose that the extremum γ is of class C^2 on the interior of its domain $[a, b]$. As we have just seen, this is the case if the Lagrangian L is regular. Then, using the Euler–Lagrange equation, we have

$$\frac{\mathrm{d}}{\mathrm{d}t} L(t, \gamma(t), \dot{\gamma}(t)) = \partial_1 L(t, \gamma(t), \dot{\gamma}(t)) + \partial_2 L\big(t, \gamma(t), \dot{\gamma}(t)\big)\dot{\gamma}(t)$$

$$+ \partial_3 L\big(t, \gamma(t), \dot{\gamma}(t)\big)\ddot{\gamma}(t)$$

$$= \partial_1 L(t, \gamma(t), \dot{\gamma}(t)) + \frac{\mathrm{d}}{\mathrm{d}t} \partial_3 L\big(t, \gamma(t), \dot{\gamma}(t)\big)\dot{\gamma}(t)$$

$$+ \partial_3 L\big(t, \gamma(t), \dot{\gamma}(t)\big)\ddot{\gamma}(t)$$

$$= \partial_1 L(t, \gamma(t), \dot{\gamma}(t)) + \frac{\mathrm{d}}{\mathrm{d}t}\Big(\partial_3 L\big(t, \gamma(t), \dot{\gamma}(t)\big)\dot{\gamma}(t)\Big).$$

Hence

$$\frac{\mathrm{d}}{\mathrm{d}t}\big(L(t, \gamma(t), \dot{\gamma}(t)) - \partial_3 L(t, \gamma(t), \dot{\gamma}(t))\dot{\gamma}(t)\big) = \partial_1 L(t, \gamma(t), \dot{\gamma}(t)).$$

If we now fix $c \in (a, b)$ and integrate, we obtain

$$L(s, \gamma(s), \dot{\gamma}(s)) - \partial_3 L(s, \gamma(s), \dot{\gamma}(s))\dot{\gamma}(t) = \int_c^s \partial_1 L(t, \gamma(t), \dot{\gamma}(t))\mathrm{d}t + d,$$

where d is a constant and $s \in (a, b)$. This equation is called the *second Euler–Lagrange equation*. In particular, if L is independent of t and $\tilde{L}(x, y) = L(t, x, y)$, then

$$\tilde{L}(\gamma(s), \dot{\gamma}(s)) - \partial_2 \tilde{L}(\gamma(s), \dot{\gamma}(s))\dot{\gamma}(t) = d.$$

The constant d of course depends on the extremal. The second Euler–Lagrange equation often has a simpler form than the (first) Euler–Lagrange equation, particularly when L is independent of t, and so is easier to apply.

Example. We consider the fixed endpoint problem with $I = [-1, 1]$, Lagrangian $L(t, x, y) = x^2(1 - y)^2$, defined on $\mathbb{R} \times \mathbb{R}_+^* \times \mathbb{R}$, and endpoint values $\frac{1}{2}$ and 1. Assume that γ is a minimum. As L is of class C^2 and regular, γ is of class C^2 and so satisfies the second Euler–Lagrange equation, i.e.,

$$\gamma^2(1-\dot{\gamma})^2 + 2\gamma^2(1-\dot{\gamma})\dot{\gamma} = c,$$

which can be simplified to

$$\gamma^2\dot{\gamma}^2 = \gamma^2 - c.$$

As the left-hand side of the equation is positive, $\gamma^2 \geq c$. Also, given the endpoint values, γ cannot be constant so, at least on some open subinterval J of I, $\gamma^2 > c$. Setting $u = \gamma^2$ we obtain $\dot{u}^2 = 4(u-c)$ and on J

$$\frac{d}{dt}(u(t)-c)^{\frac{1}{2}} = \frac{1}{2}\frac{\dot{u}(t)}{(u(t)-c)^{\frac{1}{2}}} = \pm 1.$$

Thus there is a constant b such that $u(t) = (t+b)^2 + c$. We claim that there are adjacent open intervals K and J such that $u(t) = c$ on K and $u(t) > c$ on J. If this is not the case, then from the continuity of $\dot{u}$, we have $u(t) = (t+b)^2 + c$ on I. Now, using the end-point conditions, we obtain

$$b = \frac{1}{8}, \ c = -\frac{17}{64} \implies u(t) = \left(t+\frac{1}{8}\right)^2 - \frac{17}{64} = t^2 + \frac{1}{4}t - \frac{1}{4}.$$

As $u(0) = -\frac{1}{4}$, u is negative on an open interval containing 0, which is impossible. This establishes our claim. On K, $\ddot{u} = 0$ and on J, $\ddot{u} = 2$. Therefore u is not of class C^2 and hence γ is not of class C^2, a contradiction. It follows that there is no minimum.

In this chapter we have only given an introduction to the calculus of variations, concentrating on one particular problem, the fixed endpoint problem. There is a lot more to be said, even about this problem, and of course there are other types of problems, for example, with different constraints. Many problems require more advanced techniques. There are various works written on the subject. I would suggest looking first at the books by Troutman [25] and Sagan [21] and then that of Dacorogna [10]. More advanced texts are those of Morse [17] and Struwe [24]. Troutman's book gives many applications of the theory.

References

1. Avez, A.: Differential Calculus. J. Wiley and Sons Ltd, New York (1986)
2. Barvinok, A.: A Course in Convexity. American Mathematical Society, Providence, RI (2002)
3. Beauzamy, B.: Introduction to Banach Spaces and their Geometry. North-Holland, Amsterdam (1983)
4. Bonsall, F.F., Duncan, J.: Complete Normed Spaces. Springer-Verlag, Berlin (1970)
5. Borwein, J., Lewis, A.S.: Convex Analysis and Nonlinear Optimization, 2nd edn. Canadian Mathematical Society, Vancouver BC (2006)
6. Cartan, H.: Differential Calculus. Kershaw Pub. Co., London (1983)
7. Chojnacki, W.: On Banach algebras which are Hilbert spaces. Ann. Soc. Math. Pol. Ser. I Comment. Mat. Prac. Mat. **20**, 279–281 (1978)
8. Cox, D., Little, J., O'Shea, D.: Ideals, Varieties and Algorithms, 2nd edn. Springer-Verlag, Berlin (1997)
9. Cohn, P.M.: An Introduction to Ring Theory. Springer-Verlag, Berlin (2000)
10. Dacorogna, B.: Introduction to the Calculus of Variations. Imperial College Press, London (2004)
11. Goldman, L.: Integrals of multinomial systems of ordinary differential equations. J. Pure Appl. Algebra **45**, 225–240 (1987)
12. Grünbaum, G., Ziegler, G.: Convex Polytopes, 2nd edn. Springer-Verlag, Berlin (2003)
13. Hall, G.R., Meyer, K.R.: Introduction to Hamiltonian Dynamical Systems and the N-Body Problem. Springer-Verlag, Berlin (1992)
14. Hofer, H., Zehnder, E.: Symplectic Invariants and Hamiltonian Dynamics. Birkhauser, Basel (1994)
15. Libermann, P., Marle, C., Tamboer, A.: Symplectic Geometry and Analytical Mechanics. Springer-Verlag, Berlin (2003)
16. Man, Y.K.: First integrals of autonomous systems of differential equations and the Prelle-Singer procedure. J. Phys. A: Math. Gen. **27**, 329–332 (1994)
17. Morse, M.: Variational Analysis. Wiley, New York (1973)
18. Prelle, M.J., Singer, M.F.: Elementary first integrals of differential equations. Trans. Amer. Math. Soc. **279**, 215–229 (1983)
19. Rotman, J.: Galois Theory, 2nd edn. Springer-Verlag, Berlin (1998)
20. Rudin, W.: Functional Analysis. McGraw-Hill, New York (1991)
21. Sagan, H.: Introduction to the Calculus of Variations. Dover, Weinstock (1992)
22. Schneider, R.: Convex Bodies: The Brunn-Minkowski Theory. Cambridge University Press, Cambridge (1993)

R. Coleman, *Calculus on Normed Vector Spaces*, Universitext,
DOI 10.1007/978-1-4614-3894-6, © Springer Science+Business Media New York 2012

23. Strelcyn, J.-M., Wojciechowski, S.: A method of finding integrals for three-dimensional dynamical systems. Phys. Lett. A **133**, 207–212 (1988)
24. Struwe, M.: Variational Methods, 4th edn. Springer-Verlag, Berlin (2008)
25. Troutman, J. L.: Variational Calculus and Optimal Control, 2nd edn. Springer-Verlag, Berlin (1996)

Index

R. Coleman, *Calculus on Normed Vector Spaces*, Universitext,
DOI 10.1007/978-1-4614-3894-6, © Springer Science+Business Media New York 2012